Risk Management and FMEA for Product Development

Your illustrated guide to reducing time-to-market through risk management and FMEA

4th Edition

By: José Campos

Published by:

Rapidinnovation Press
www.Rapidinnovation.com

Risk Management and FMEA for Product Development (4th Edition)

By: José Campos

Published by:
Rapidinnovation Press
Rapidinnovation, LLC
25000 SW Big Fir Rd.
West Linn, Oregon 97068
United States
www.rapidinnovation.com

ISBN-13: 978-0-9817595-0-0
ISBN-10: 0981759505

LCCN: 2012913573

Published in the United States

CIP data available from the publisher

Get your interactive PDF version FREE!
See the last pages of this guidebook for details

V-RMFMEA/082712/6.2.4/ACS/PRN/SW

Introduction

Legend 7
Foreword. 8
Approach 9
About the Author 11

Critical Definitions and Concepts

What is Risk

Introduction. 13
The System for Closed-Loop Kaizen 15
Types of Risk. 16

Why Risk Management

The Business Case for Risk Management 17
Quantitative and Qualitative Risk Management. 19
Risk in Product Development 19
Process Map: Risk and the Product Life Cycle 21
Understanding Risk Management 23
Critical Success Factors (CSFs). 25
The Uncertainty Continuum of Risk 26
Proactive Risk Management (PRM). 30

Roadmap to Maturity

Risk Maturity 31
Risk Management in Your Organization's Culture (DNA) 32
Selection of Risk Methods Using the Complexity Diagram 34
Types of Risk Management to be Performed. 38
Example: Types of Risk Management 39

Teams and Team Charters

Introduction: Risk Management Teams 40
The Skill Set 42
Why Risk Management Efforts Fail 43
The Risk Management Office (RMO) 45
Charter for a Risk Management Team 47

Risk Management

Overview of Risk Management

Risk Management 50
Detailed Process: Risk Management 52

Introduction

Identifying Risks

Protocol for Risk Definition . . . 58
Examples: Using Yellow Sticky to Document Risk . . . 60

Prioritizing Risks

Prioritizing Risks: Creating a List of the Vital Few . . . 65
Examples: Karnaugh Map 2x2 Matrixes . . . 66

Planning Preventive and Corrective Action

The Risk Management Plan . . . 67
Risk Management Plan (RMP) as a Spreadsheet . . . 69
Documenting Your Risk Management Plan . . . 71

Implementing Preventive Action

Plan for Success and Failure . . . 72

Monitoring the Risks

Proactive Risk Management . . . 73
Monitor Your Risk Plan Using the Critical Questions of Proactivity . . . 74

Taking Remedial or Corrective Action

Taking the Next Steps . . . 75

FMEA: Failure Mode and Effect

Using FMEA in Risk Management

FMEA: A Framework of Risk Management . . . 77
The Business Case for FMEA . . . 78
Types Of FMEA . . . 79
The Attributes of FMEA . . . 80
Risk Management Strategy Matrix for a Development Program . . . 82
FMEA Glossary . . . 84
Detailed Process: FMEA . . . 86
The FMEA Management Plan . . . 94
Examples of Ranking . . . 97

Critical Tools

Deployment of a New Tool

Cascading a New Tool to the Organization . . . 107
Detailed Process: Deploying a New Tool . . . 109
Planning Checklist . . . 111
Planning Document . . . 112

Tips for Successful Deployment . . . 113

Problem-Solving Process

Problem Solving Process . . . 114
Detailed Process: Problem Solving . . . 115
Problem Solving . . . 118

Karnaugh Maps: 2x2 Matrixes

Prioritize Items from a Selection of Many . . . 119
Example: Karnaugh Map (2x2 Matrix) . . . 121
Pair-Wise Comparison . . . 122
Detailed Process: Prioritizing a Karnaugh Map . . . 123

The 5 Whys

The 5 Whys . . . 125
Examples: The 5 Whys . . . 126

Yellow Sticky Protocol

Effective Group Participation . . . 127
Level 3 Detailed Process: Yellow Sticky Protocol . . . 128
Example: Yellow Sticky Protocol . . . 129

Affinity Diagrams

Organize Information and Reach Consensus . . . 130
High-Level Process . . . 131
Detailed Process: Affinity Diagrams . . . 132

Prioritizing Affinity Diagrams

The Process to Prioritize . . . 134
Level 3 Detailed Process: Prioritizing Affinity Diagrams . . . 135
Example: Affinity Diagram . . . 136

The Fishbone Diagram

The Fishbone Diagram . . . 137
Detailed Process: Fishbone Cause and Effect . . . 139
Developing Scales for Risk Management . . . 142

Appendixes

Appendix A: Recommended Books

Introduction . . . 145
Teams and Teamwork for Product Development . . . 145
General Management and Strategy . . . 146
Product Development . . . 146
Productivity Improvement . . . 147

Introduction

Project and Program Management. 148
The Voice of the Customer (VOC) 149
Risk Management and FMEA (FMCA) 151

Appendix B: Glossary

Definitions for Product Development 152

Appendix C: FMEA Process Map from SEMATECH, Inc.

Overview. 175

Appendix D: Templates

Instructions 176
Types of Risk Management to be Performed. 177
Charter for a Risk Management Team 178
Documenting Your Risk Management Plan. 180
Examples: Documenting Your Risk Management Plan 183
Monitor Your Risk Plan Using the Critical Questions of Proactivity 184
Risk Management Strategy Matrix for a Development Program 185
FMEA Management Plan 187
Planning Checklist for Deploying a New Tool 189
Planning Document for Deploying a New Tool 190

Get Your Interactive Version FREE! . . . . *details in the back*

Other Titles You May Be Interested in Owning . . . . *details in the back*

For comments, questions or additional copies please e-mail the author: risk@rapidinnovation.com.

Legend

Navigate through this Guidebook

To ease navigation, we have provided a set of icons, which provide orientation and clarity. They are consistent throughout the guidebook, always carrying the same meaning.

Symbols to Indicate Facts to Know

Important Suggestion

Used with each important suggestion, note, or tip, derived from experience in the field during actual projects.

Essential Quote

A quick way to read very important information. We suggest you read all essential quotes.

Flexibility

Indicates the flexibility of a particular tool. We encourage you to adapt the tool to your specific situation and not be governed by rigid, prescriptive frameworks.

Definition

A defined term. See also our Glossary at the back of the guidebook.

Time Frame

To help you plan the likely duration of a meeting, work session, or other event.

You Are Here

One more way to get oriented or to check that you have take the proper steps up to this point.

Symbols for Links to Helpful Info

Note for Print Version

Interactive features are available in the electronic PDF version. If you have a printed version, contact the authors to learn how to upgrade your guidebook.

Template (Printed Version)

You will find copies of the templates at the back of your guidebook.

Template (Electronic Version)

Double click this icon to open a corresponding template (enabled for use in Adobe Reader and Acrobat) that you can fill with your own information and share with your peers and colleagues. When you close the template, you will return to this guidebook.

Each time you open a template from your guidebook, it will be fresh. But you can save versions that you have completed, partially filled in or left blank for future access. Saved versions remain enabled for revisions and collaboration.

These enabled templates also provide you with Comment & Mark-up Tools. Just as you might with a hard copy, you and your team can highlight text, add notes and make custom annotations.

See Also (Electronic Version)

Clicking this icon will jump you to a place in the guidebook with more information relevant to the topic at hand.

Back Button (Electronic Version)

After following a link, you can easily click to return to your original place in the guidebook.

Introduction

Foreword

By Preston G. Smith

I am greatly pleased to see this workbook. José Campos has done a wonderful job of presenting this material in a way that makes project risk management straightforward to assimilate. The book needs no introduction by me, so I will make a few observations that may help you to place it in context.

Based on years of experience in teaching risk management to project groups in North America, Europe, and Asia, I firmly believe that project risk management is a skill that is learned only through practice on actual risks. Consequently, you will pick up these skills fastest by choosing some risks from a project you are working on and running them through the steps suggested herein. This way, you see where judgment is needed, and you start to acquire this judgment. What looks easy while just reading a page may turn out not to be so easy!

You can extend this learning by saving the homework you do to see how the risks you have managed as exercises actually work out in reality. Did you describe your risks in sufficiently clear terms that, looking back, you can see now that you were describing actual threats to your project. (One of the skills learned by newcomers is describing the risk distinctly enough initially so that effective mitigation plans can be formulated.) Did the risks you choose to manage turn out to be the important ones, or did others eclipse it? Did the mitigation plans you created fit with what really happened on the project, or were they fantasy?

Another observation I have made in watching many professionals attempt to deal with the risks on their projects is that they do very well on the initial step (identifying their project's risks), but then they drop the ball. The subsequent steps somehow don't happen, either because the team loses interest or momentum or because it becomes so overwhelmed with the total risk picture that it wants to deny its risk list. This is worse than no risk management, because when the risks on this list start happening, it is embarrassing that the risks were known but no action was taken.

Remember that all of the benefit in project risk management comes in the last step — carrying your mitigation plans through so that the risk either does not happen or is less severe than it would have been. If you find that you don't have enough time to do as complete a job of risk management as you would like, don't put all your effort into the identification step and stop. Instead, do a brief identification and follow the identified risks through to effective mitigation planning. Then you have something of value, albeit perhaps for only a subset of the risks you face.

This brings me to a final point: How do you know how much time to put into risk management? Admittedly, risk management is a time-consuming activity, and you can do too much of it, thus wasting valuable project time. The answer is a balance between the effort that goes into it and payoff of not having certain project risks occur. You find the proper balance only by managing the risk in some projects and watching the outcome. Keep metrics on the amount of project time or effort that is saved by risks that do not happen, and compare this to what you are spending on risk management. Then increase or decrease your risk management effort until the effort pays for itself.

Best wishes as you proceed!

Preston G. Smith
Coauthor, *Proactive Risk Management* (Productivity Press, 2002)

Approach

Why We Wrote this Guidebook

Every product development project has plenty of risks — things that could go wrong. Risk is always there; your only choices are to ignore it or to manage it. Regrettably, too many organizations choose to ignore risks, perhaps hoping that they will miraculously disappear. But risks have a way of becoming reality, disrupting and even destroying good development projects. We felt that if we provided a set of tools, templates, examples and processes on risk management, organizations would be more inclined to manage risk.

A How-To Manual

Having been on the receiving end of guidebooks, handbooks, "primers" and training during our lives in the corporate world, we realized that often the information was not sufficient for us to actually use it in our situation, or the information was formatted poorly such that it made it impossible for us to apply it, which resulted in much frustration, not to mention loss of productivity. Worse yet, the documentation we received from training sessions was simply a hard copy of the PowerPoint slides shown during the session, which were of little help to us when it came time to apply the learning.

We made a commitment that our guidebook would contain everything a professional would need to apply every concept in it. We have strived to provide the tools, methods and encouragement for you to successfully understand and apply the voice of your customers.

How to Use this Guidebook

We developed this guide for individuals and teams who are in product development (Marketing and R&D) and have specific assignments to deliver results. The entire focus of the guidebook is to lead you — step-by-step — through each and every concept, even if you have never performed the task before.

The sequence of this guidebook is based on a chronological approach — the outline and format follows a logical "starting point" and move towards a "end point." This means that the higher the page number, the later in the process.

This is not a book about concepts. We have limited concepts to the minimum necessary to create context or perspective. The focus of this guidebook is on the "how-to."

The Process Maps

The most important element of this guide is the process maps — the step-by-step notes for starting and completing key tasks. These process maps lead to the various tools and templates.

There is a hierarchy of process maps:

LEVEL 1: THE MASTER PROCESS This is the highest view of the process. The Master Processes, also known as Mental Models, represent big pictures, and they are rarely actionable due to their strategic nature. They are simply here to give you the overall perspective of a large endeavor or system.

LEVEL 2: MAIN PROCESS These are the major phases contained in the guidebook. They provide a phased (stage) approach and are indeed actionable. Main Processes as generally good for planning purposes, also to discuss with your team as you prepare to start a particular stage.

LEVEL 3: DETAILED PROCESS These are the processes you will need to start and complete tasks. They not only contain the steps, but we have also included critical information in the form of tips, suggestions to help you locate the progress and other important information, please see the Legend at the start of this guidebook for further details on the icons used to help you navigate.

Introduction

The Template and Tools

Templates are forms that contain information that you and your team enter in order to collaborate and create archival information for later use. Tools are matrices, spreadsheets, tables and other instruments.

It is important to note that for every tool or template there is at least one example and in some cases several. This is intended to provide you with a general idea of the information that should be entered. It is not intended to limit what you and your team feel needs to be done.

Unique Capability for the Interactive PDF Version

Not only does every template in this book have a full example, but also it is fillable. You and your team can enter information about your project and share it. You will notice a link in the form of an icon and when you click on it, you will be able to open a blank template that you and your team can use. Once you have completed entering the information, the template becomes its own PDF independent of the guidebook. You can save it using your own name or title, which you can also share with the rest of your team around the world. You can continue entering information until you are satisfied.

Flexibility

We understand that every development project is different — by definition. Consequently, our intent is not to limit your flexibility. All the process maps, templates and tools can be modified to better meet your objectives and style. We suggest that the first time you apply the these process maps, templates and tools that you try to adhere to our format. As you become familiar with the intent, you can modify them until you have something that is unique to your organization and also meets your requirements to the fullest. Don't be afraid to experiment and modify.

Essential

"The most important element of this guidebook is the process maps — the step-by-step notes for starting and completing key tasks."

Mission of this Guidebook

The mission is to help you and your team implement a model of Risk Management that is repeatable and sustainable such that you reduce time-to-market and improve the quality of your new products.

Risk Management is a powerful way to improve quality, but you must be able to do it on an on-going basis; otherwise, you will not derive all the benefits. This guidebook was written to help you adopt Risk Management as an integral part of your product development model.

Objectives of this Guidebook

The overall objective is to create the capacity for improved productivity! And the results of using this guidebook will include these specific objectives:

- You will be able to you reduce or eliminate the impact of uncertainty (risks) on your product development projects.
- You will be familiar with key strategies needed for effective risk management.
- You will be able to apply the material presented in this guidebook to real-world situations.

Your Input

No book is perfect; this one is no exception. If you find difficulties in using the tools, concepts and processes, do not hesitate to contact the authors: risk@rapidinnovation.com.

About the Author

José Campos, Principal Consultant and Founder, Rapidinnovation

José collaborates with B2B technology companies around the world on the methods, tools and cultural shifts to rapidly and consistently deliver superior value to customers.

José has spent his career working to define and deliver innovative products to market. For 30 years he has passionately focused on discovering new methods and approaches to discover and deliver superior solutions to the customer in the tumultuous world of high-tech

He has written several acclaimed books: *Voice of the Customer for Product Development* — a methodology to discover, prioritize, and use customer requirements to create clearly differentiated products — faster than the competition. Also, *Risk Management for Product Development* — the strategies and methods to reduce the impact of uncertainty on your investment. Jose was coauthor of Project Management Toolbox in collaboration with Dr. Dragan Milosevic.

His focus on high-tech and his technical background allow him to provide meaningful and practical coaching to technology professionals. He has traveled the world help R&D and Marketing teams improve their ability to create innovative products that customers love.

Related Guidebooks for Product Development

In addition to *Risk Management and FMEA for Product Development*, our library includes several related guidebooks in the Product Development series:

- The Voice of the Customer for Product Development
  Your illustrated guide to obtaining, prioritizing and using customer requirements and creating winning products.
- Actionable Metrics for Product Development
  Your illustrated guide to developing and using metrics to improve product margins and reduce time-to-market.
- Debriefs and Postmortems for Product Development
  Your illustrated guide to improving performance through lessons learned.
- Flexible Project Management for Product Development
  Your illustrated guide to making project management work in fast-changing environments.

Training, Consulting and Facilitation

Our highly experienced authors are also available to provide direct consulting to your organization. They can be team facilitators to help you implement the contents of these guidebooks. They can provide training on the topic plus many other areas of product development. Contact us for more information.

For comments, questions or additional copies please e-mail us: jose@rapidinnovation.com.
Visit us at www.rapidinnovation.com.

Get Your Interactive PDF Version FREE!

See the last pages of this guidebook for details on how to acquire your free copy of the interactive PDF version.

PRODUCT DEVELOPMENT

Critical Definitions and Concepts

What is Risk 13
Why Risk Management 17
Roadmap to Maturity 31
Teams and Team Charters .40

What is Risk

Introduction

Risk is the chance that something can go wrong with your project that will cause a negative impact (loss, damage, etc.). Corollary: Ignoring risks does not make them go away! On the other hand, risk management is the disciplined process of identifying risks, evaluating them, preparing a response and implementing the response at the right time to avoid or reduce damage to your project. In short, risk management is answering the following key questions:

1. What can go wrong?
2. What is the probability of it happening?
3. What would the impact or consequences be if it does happen?
4. What can we do to avoid it happening?
5. What can we do if it does happen?

Positive Risks?

The other side of risk is Critical Success Factors (CSFs), which are those things that must happen to deliver the expected results. While this guidebook focuses on risk, we also include a small section about CSFs.

We highly recommend that you address both categories in your projects. The methodologies and processes are the same except that for risks you focus on things that *could* go wrong and *must not* happen, while with CSFs you focus on those few things that *must* happen.

The Compound Murphy's Law

- If something can go wrong, it will.
- Even if it cannot go wrong, it will.
- Either way, your manager will be there to see it when it happens.
- And it will happen the day before your annual performance review.

Essential

"If you are not managing risk, you are not managing your project."
—Dragan Milosevic, PhD

ISO-31000:2009

The International Standards Organization (ISO), based in Switzerland, issued a new standard to manage risk in an enterprise called: *Risk management — Principles and Guidelines, Edition 1*. As of this writing (January 2011) the standard was not for certification.

The standard defines risk as: The Effects of Uncertainty on Objectives. This guidebook indeed adheres to the strategy of the ISO standard. The authors enthusiastically support the issuing of a new standard for manage risk.

For additional information please visit: http://www.iso.org. Then search for ISO-31000.

Proactive Risk Management (PRM)

In this guidebook we stress the need for Proactive Risk Management (PRM). Some organizations have felt that developing a plan was all you needed to do to ensure good management of the risks. But this does not work because the plan is a static document with no life in it; i.e., it is a passive approach.

PRM does not mean that your plan is stored in a filing cabinet or a hard drive, never to be seen again. Rather, PRM means you are constantly monitoring and updating your plan. It also means your mitigation steps are always up to date, in step with the changes in your project. Above all, it means an attitude of proactivity — the realization that it is your responsibility to keep the plan alive and to take appropriate action without having to be asked.

What is Risk

Good Indicators

Some of the good indicators that you are practicing Proactive Risk Management include:

- Designating a risk manager who is fully aware of the responsibilities of this role and has the skills and abilities to serve in this capacity.
- Chartering a risk management team that meets regularly to monitor the plan, update mitigation and anticipate any disruptions.
- Having a management sponsor who is eager to help, who proactively ask for updates and ways to help the team succeed.
- Writing a Risk Management Plan that is continually updated, is conveniently available to all team members and contains clear, action-oriented mitigation steps.
- Ensuring that the entire development team, managers and supervisors clearly understand the reason for risk management.

The System for Closed-Loop Kaizen

Continuous Improvement at Its Best

This diagram shows the opportunities of combining Risk management with debriefs. Risk management is about predicting what could go wrong with your project, and debriefs are about analyzing what went right and wrong with your project. It stands to reason that the two should be connected. Think of them as the bookends for every project.

> **Important**
>
> **Assess the Effectiveness of Risk Management**
>
> Applying a model of risk management to your development model is a very good first step. Next, you must assess if the model you have implemented is improving your development ability. We highly recommend that you conduct debriefs (also called postmortems and retrospectives) on your development project, with the specific objective of evaluating the risk management model.

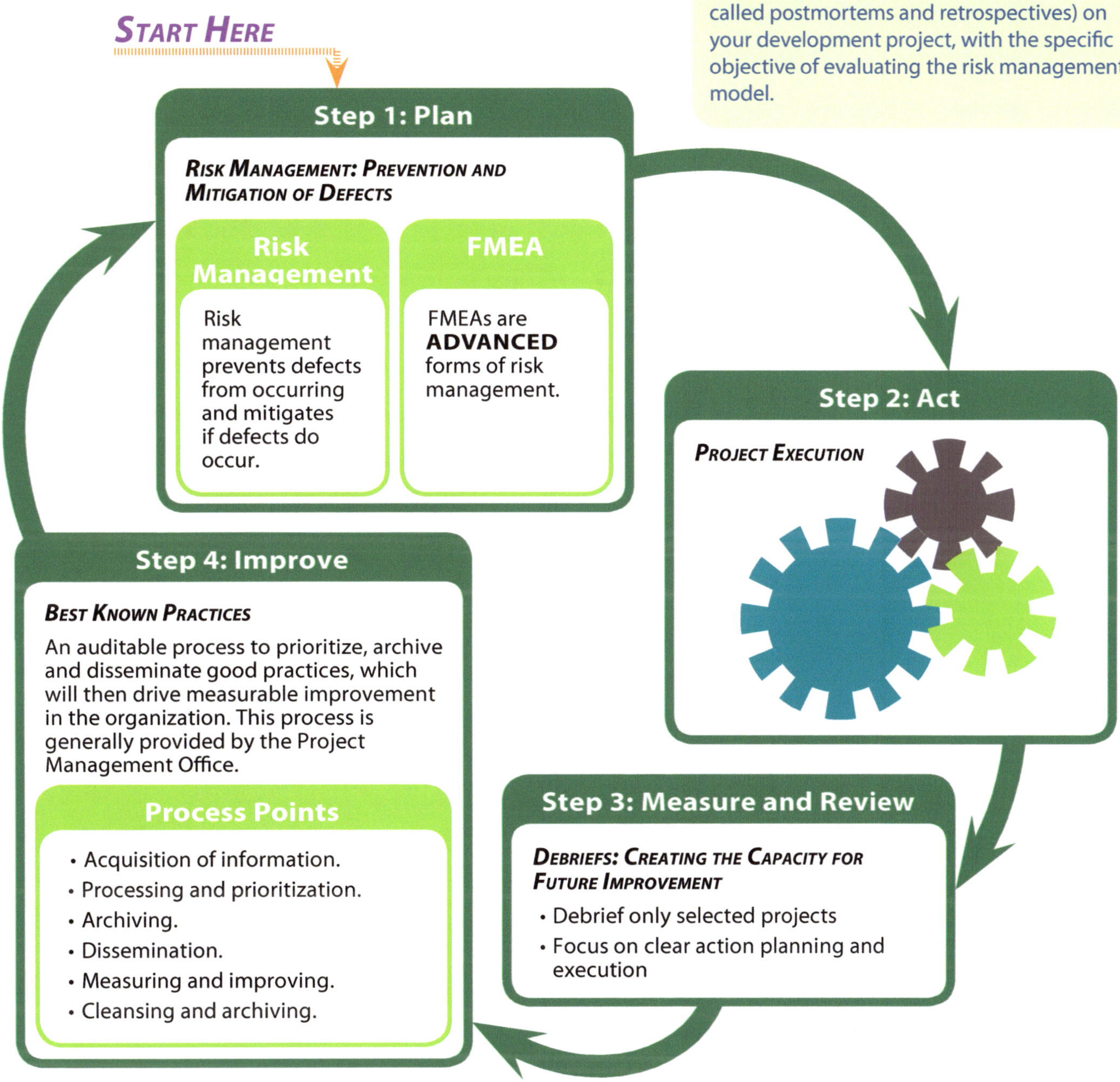

Implement the System for Closed-Loop Velocity Improvement

What is Risk

Types of Risk

A Few Examples

There are many types of risks. We encourage you to look at the big picture of your project and avoid tunnel vision. Risks include everything that might interfere with the delivery of the objectives of your project. Some risks will be unique to your industry, your company and even the country in which you reside. Use the following sample as a starting point, continually improving it with risks more relevant to your situation.

Types of Risk	Examples
1. Political (Internal and External)	• Personal agendas create obstacles to your project. • Two managers compete for a promotion; your project is caught in the middle. • Government or industry regulatory mandates interrupt your project. • New safety regulations cause a long delay for your project.
2. Cultural	• Your project involves a cultural change in the organization, causing obstacles. • A new acquisition means you must deal with two (corporate) cultures: that of your own company and of the new company. • Conflict arises because your team is scattered in across several countries. • The CEO of the company is risk-averse and considers your project too risky, while you believe it to be very manageable.
3. Economic	• A recession causes budget cuts and layoffs. • Added costs make your project unprofitable.
4. Technical	• The technology you are using in your project does not work. • The technology you are using creates a safety hazard.
5. Legal	• The owner of the IP you are using sues your company for infringement. • A restrictive law in another country will impact the IP you are planning to use.
6. Weather	• A sudden snowstorm brings your project to a halt. • An earthquake damaged the facilities of one of your suppliers.
7. Resources	• Your key engineer takes ill and is unable to continue working. • Two of your top scientists are lured away by a competitor. • A layoff could reduce your team.
8. Organizational	• A reorganization is announced in the middle of your project, causing you to lose several engineers to a new division. • Your division has no method to prioritize projects — every project is "Priority 1." • Your company is acquired and a new culture emerges, thus creating disruptions.
9. Procurement	• Technology you wish to purchase is unprotected and available to competitors. • The factory that you wish to use to manufacture your new product has gone on strike and there is no end in sight.
10. Market	• A new industry standard is enforced, which may cause a product redesign. • A new technology is being adopted by your competitors, which may cause redesign or a complete new design for your products.

Why Risk Management

The Business Case for Risk Management

According to ISO-31000:2009 (Risk Management — Principles and Guidelines), risk is the effect of uncertainty on objectives. We can paraphrase the ISO standard and conclude that risk is the probability that something could go wrong with your project, which would interfere with obtaining the stated business objectives. Furthermore, we can conclude that risk management is all about reducing or eliminating the uncertainty that might interfere or negate business objectives.

We encourage you to factor risk into the business model of your organization and resist the temptation to see risk as yet another pro forma, "boilerplate," requirement. At the end of the day, it's not about the management of risk; it's all about the achievement of objectives.

Four Goals

The business case for deploying a framework of risk management in product development is to achieve any of four specific goals:

1. **REDUCE TIME-TO-MARKET.**
2. **IMPROVE RELIABILITY.**
3. **REDUCE WASTE (IMPROVE EFFICIENCY).**
4. **IMPROVE CUSTOMER SATISFACTION.**

Essential: "Risk is the effect of uncertainty on objectives." — ISO-31000:2009 Risk Management Principles and Guidelines.

The need to improve productivity, as defined by the four items above, is an enduring effort of any product development company, particularly those that face global competition. Yet risk management is rarely seen as an area of productivity improvement; thus, it is one that has not been exploited. But risk management is a way to improve efficiency in product development. Furthermore, risk management has a direct and clear impact on profits. For these reasons, we suggest that risk management should be part of your business model.

The diagram on the right defines several business benefits of risk management. In its simplest form, the business case for risk management can be made by the following assertion: Risk is always there; ignoring it is a flawed strategy that will not make risk go away.

How many times have you heard the following statement: "That problem came out of nowhere."? We beg to differ; most "surprises" in product development are predictable and can be mitigated before they become a big problem.

The challenge in product development is time-to-market. Risks that are not managed have a high certainty of extending your development time. They disrupt the development process and cause your resources to address problems that should never have happened. Consequently, one way to reduce time-to-market is to have a rigorous risk management model.

Further, reliability is improved because problem areas are identified early enough to allow for preventive action or changing the design at the point where it costs the least — i.e., the start of the development project.

Finally, customer satisfaction is improved because your new product is more reliable — service and support will be enhanced through risk management.

The Benefits of Risk Management: Increased Probability of Achieving Your Business Objectives

Why Risk Management

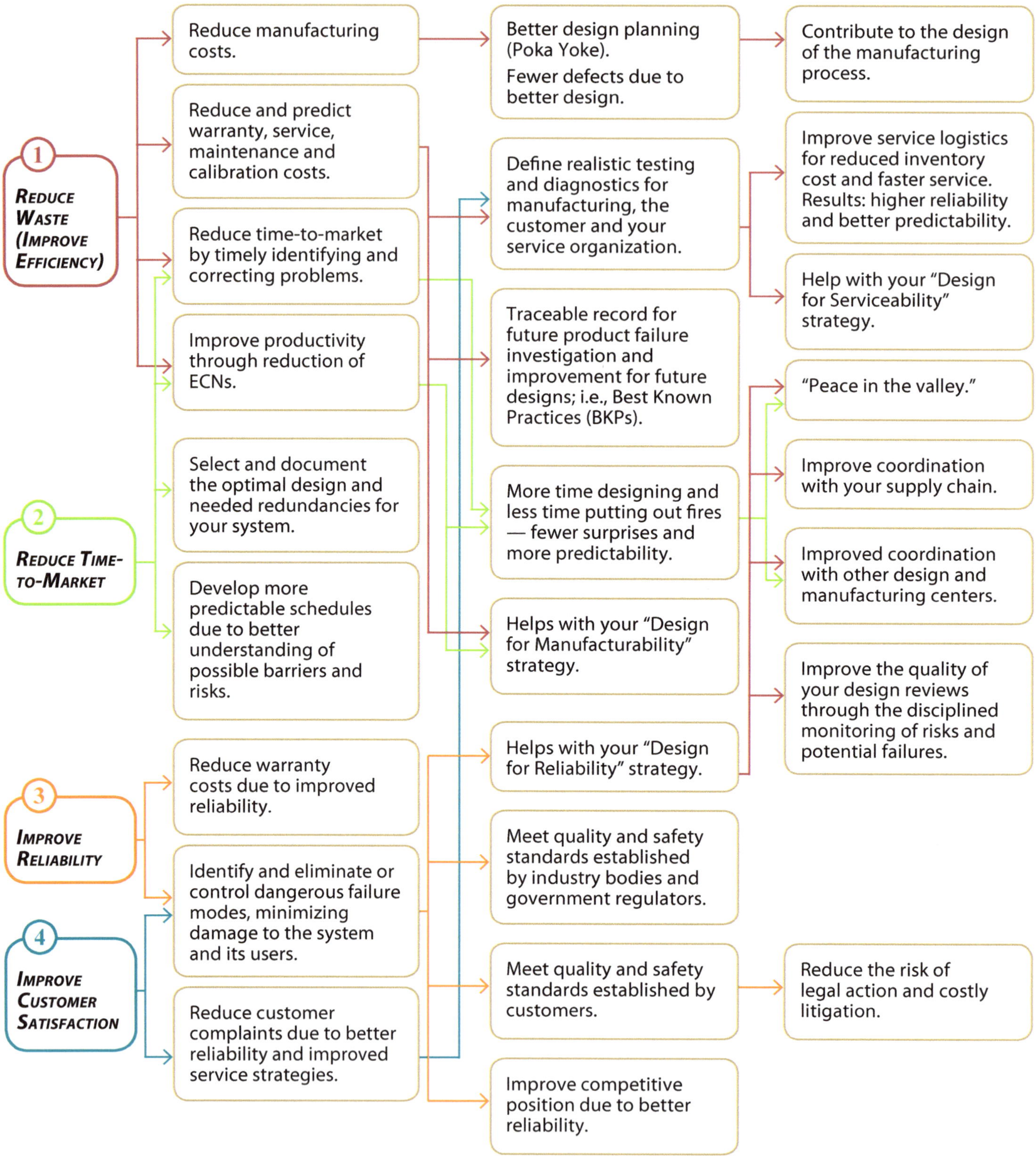

The above diagram shows the benefits of risk management with more specificity. The left column is the original premise as outlined on the previous page. Quantification of the benefits varies from company to company, and from project to project, but we hope the diagram helps you quantify your situation.

Quantitative and Qualitative Risk Management

Generally the current methods of risk management for product development tend to be the qualitative types, where you rely on the experience of the participants to identify and then prioritize risks. Linear scales from 1 to 10 are used simply to gain velocity during the process. There is a threat in relying solely on qualitative, linear scales; they may bear no relevance to reality. Consider that you are depending on the opinion of a group of people, which means "subjectivity."

There is a trade-off between quantitative (complexity) and qualitative (opinion) risk management. Quantitative will give you better and more reliable results over time, which will create a higher degree of credibility. Qualitative risk management provides speed and ease of use.

Here are some suggestions to help you with this very important trade-offs:

- **YOUR TEAM HAS NEVER DONE FORMAL RISK MANAGEMENT:** We highly recommend that you start with the qualitative method shown in this guidebook. Then, start building a database of the scales you use and set about improving the meaning (relevance) over time by constantly refining the criteria and statistical validity. We suggest that you also perform debriefs (postmortems) on projects where you used risk management to capture Good Practices and ways to improve the accuracy of your scales so future projects benefit from your learning.
- **YOUR TEAM HAS BEEN DOING FORMAL RISK MANAGEMENT BUT STILL QUALITATIVE:** We suggest that you start doing debriefs in order to determine the efficacy of your risk management model, capture data to improve the statistical validity and document Good Practices.
- **YOUR TEAM HAS BEEN DOING FORMAL RISK MANAGEMENT:** You may wish to apply Monte Carlo Analysis to improve the statistical validity of your risk model. We do not cover Monte Carlo Analysis in this version of this guidebook, but there is much information available via the Web, books, papers and other publications. You can get general information at www.wikipedia.com search under Monte Carlo method.
- **YOUR TEAM IS AIMING TO IMPROVE THE QUALITY OF YOUR RISK MANAGEMENT MODEL:** Form a partnership with Reliability Engineering, if you are fortunate to work for a company with such a group. Reliability Engineering employs a very quantitative form of risk management, with the focus on the components, mainly hardware. Nevertheless, they can help you initiate your efforts to improve the statistical accuracy of your model.

Risk in Product Development

As with any other area of focus, product development carries a number of differences from the traditional perspective of risk management. The traditional risk management models have been inspired by PMBOK (Project Management Body of Knowledge), which is published by the PMI (Project Management Institute). One of the roots of PMBOK lie with the construction industry, which is very different from product development, particularly in high-technology. The main difference is in the nature of the two industries — construction is mature and well-established, but product development thrives on uncertainty where the technology, products, markets, competition and personnel are changing with regularity.

Fortunately, the ways and means of risk management are scalable to address either mature or early-adopter industries. Yet the trap is to use one approach when the other should be applied. We strongly recommend that you recognize the unique characteristics of your situation and apply the appropriate risk management approach; for example:

- Is your industry stable (mature), or is it undergoing considerable change?
- Are your products simply an evolution of previous generations, or is the technology changing the paradigm?

Why Risk Management

- Are you under considerable competition, or are your competitors less aggressive?
- Is the cultural identity of your company or organization that of a risk-taker, or is it risk-averse?

The Risk-Averse and the Risk-Taker

All individuals and organizations have a preferred level of risk-taking. The project manager must be wise and identify the risk index of his or her organization. A particular risk strategy may not work because the culture of the organization may not tolerate it. On the other hand, in a risk-taking organization you may be able to push the envelope.

Consequently, your feelings (and the feelings of the project's key stakeholders) toward risk will influence overall behavior — even in the presence of data. The bias can be positive or negative, and, in some cases, subconscious. What level of risk-taker are you?

The Risk Continuum: If your organization, managers or team members are risk-averse, they will tend to be conservative and avoid risk (the left side, as depicted by this mental model). Alternatively, those who are risk-takers will move forward, often disregarding risk entirely. Is your team risk-averse or risk-taking? It's important to know because it will influence the model of risk management you implement for your organization.

The Product Life Cycle and Risk Management

In the following pages, we detail the entire concept of the product life cycle and how it relates to risk management.

Products begin way before the first line of code or the first hardware component is designed. All products can be traced to the corporate strategy that sets the direction of the company and, thus, the types of products that will be produced. This is critical to understand, as some risks have their root cause in the strategic phases of the company; for example, addressing a new market for which your company is not prepared.

Risk management can be performed at any stage of the entire product life cycle, depending on your business objectives. Furthermore, it is not limited to engineering — marketing can and should perform risk management.

Technology management has much to benefit from risk management; for example, the selection and staging of new technologies carry considerable risks that can be properly managed through risk management.

At the executive level, every strategy has a set of risks that, if addressed through formal risk planning, would help avoid failure or, better yet, ensure the success of new strategies.

Why Risk Management

Process Map: Risk and the Product Life Cycle

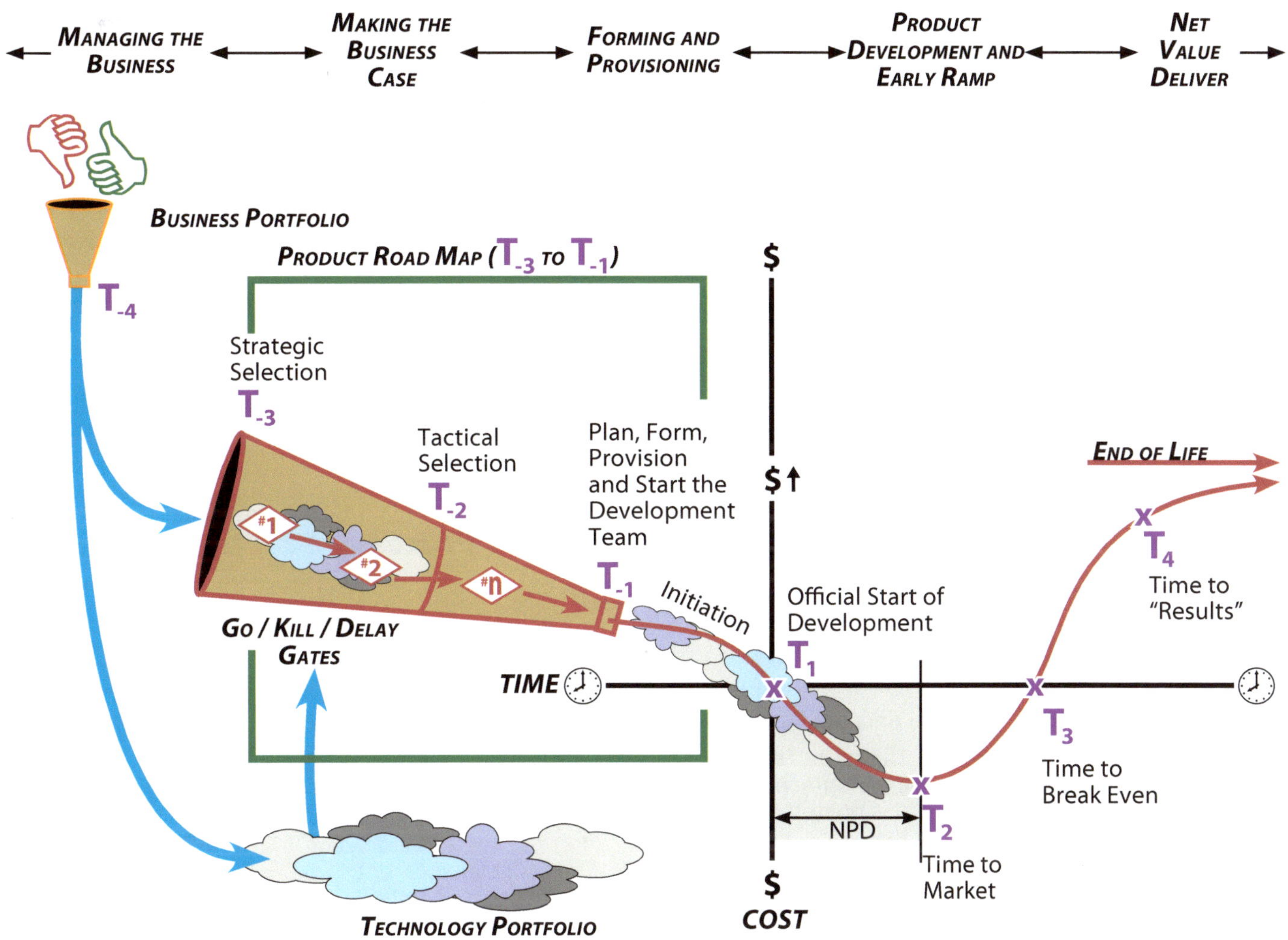

Managing the Business

- Risk in the economy.
- Macro risks.
- Risk of not understanding the marketplace.
- Risk of poor selection of technology.
- Risk of poor portfolio management.

Strategy Implementation

- Technology risks.
- Market risks.
- Risk of competition.
- Risk of a disruptive new product.

Tactical Selection of Project

- Risk of prioritizing the wrong project.
- Resource risk.

Product Development

- Risk of not understanding customer needs.
- Resource risk.
- Supplier risk.
- Manufacturing risk.
- Quality risk.
- Materials risk.
- Reliability risk.

Product Introduction

- Risk of a poor introduction to the market.
- Risk of the product not being ready to sell.
- Risk of poor support.
- Risk of a fatal bug.

Maturity

- Supplier risk.
- Manufacturing risk.
- Competitive risk.

The above mental model is a depiction of the product development system. It shows the start at point T_{-4}, where the corporate decisions are made to manage growth and profitability. Note point T_1, which shows the start of a typical product development project. Also note that the New Product Development Process (NPD) starts at T_1 and ends at T_2. You must take into consideration the entire system to develop the right risk management plan at the right time.

Why Risk Management

Risk and the Product Development Cycle

Most organizations have a process to develop new products — some call it the New Product Development (NPD) process. Whatever you call it, the process is intended to guide the development team from the start to the point where the product is ready for purchase by customers. In some cases there may be no documented process; regardless, there is a calendar-based, step-by-step progression from start to public announcement.

In addition, there are many product development methods, such as Stage Gate, Agile, Waterfall, Iterative, etc. They all require close interaction with customers.

While there are significant differences in the product development process across industries, the basic premise is common to all: the ability to develop a product on time, and at a profit. For example, software and hardware development processes are different, but the need for customer interaction is critical in both cases.

Regardless of your process or methodology, the need for risk management is always present. Ignoring risk is a flawed strategy. Also very important to know: The type of risk management varies with the phase of development. We highly recommend that you start your risk management efforts as early as possible. This approach is contrary to some conventional views that suggest waiting until the project is officially chartered to start risk management. In product development, the need starts very early indeed.

A Generalized New Product Development (NPD) Process

Start of Development — T_1

New Product Introduction — T_2

Ideation Phase

The period of time spent on understanding customer needs, also called "concept development phase."

During this phase, the focus is on managing the "unk unks" — the unknowns in the technology, the possible design and the markets.

Risk management should not interfere with the ideation phase. Properly used, risk management can be done concurrently with innovation, ideation and brainstorming.

Product Definition

The start of the development process; at this point the definition of the product is completed, but not necessarily finalized.

Risk management during this phase is more clear. Your development is now a project.

We do not recommend that you do risk management for all projects, unless you are in an industry where the technology, markets and competition are in constant flux. Rather, use your judgment and allocate your risk management resources wisely. You may choose to do risk management on an entire development program, or just parts of it.

Product Design

The new product is being designed. The definition is finalized.

You are updating your Risk Management Plan, constantly monitoring to put your plan to work should the need arise.

New risks will emerge as more information is known; this is normal, and you can accommodate the changes in your plan.

FMEA is likely being done in the engineering organization.

Final Testing and Production

At this point the product is in final refinement, and being prepared for public announcement.

In some industries, "beta" units are being tested with customers.

In this phase, manufacturing is likely to be focused on FMEA for processes.

Marketing may be doing a risk plan to better manage the new product announcement.

The above diagram is a simplified view of the product development process. Yours is much more detailed and complicated. Nevertheless, our intent is to show the role of risk management during the major phases that are common to most, if not all, product development efforts. The key points are that it is never too early to start risk management and that risk management will serve you well during the entire process.

Understanding Risk Management

Risk, Chance and Gambling

A critical step in the implementation of risk management is the clarification and adoption of a uniform language across the organization. Many of us have different perceptions or understandings of the terminology used in risk management. It is important for your organization to address these differences and develop a unified set of definitions to avoid misunderstandings.

Important! Every organization should create a one-page glossary where key risk management terms are clearly defined.

Many people have preconceived interpretations of various terms. A company glossary will go a long way toward aligning everyone to a single meaning; thus avoiding confusion.

Term	Description
1. Accuracy	• It is the degree of veracity — the closeness to an agreed standard.
2. Ambiguity	• Doubt about the meaning or lack of clarity. Something that has different meanings depending on its context. Something that can be understood in more than one way; for example, the behavior of key stakeholders in your project.
3. Chance	• Related to uncertainty, it implies the willingness to make decisions in the absence of information, i.e., "betting on the outcome." Good for gambling, but not part of managing your project. Chance is when you don't know the probability of an event.
4. Gambling	• Gambling is **NOT** the same as risk management. Gamblers can choose the risks they take. In business, the risks are there already, waiting to be managed.
5. Mitigation	• Refers to anything that counters the negative effects of the risk or the sources of the risk. It can be prevention, contingency planning, reserves, etc.
6. Precision	• The degree of reproducibility, it's related to accuracy; for example, the accuracy and precision of a measurement is established by repeatedly meaning to some known standard.
7. Predictability	• Refers to something that happens as expected. The degree to which a correct prediction can be made.
8. Repeatability	• The ability to do the same thing over and over by the same person using the same measuring devices under the same conditions.
9. Risk	• The probability of something going wrong with your project. The risk is known, and one can mitigate it, avoid it or remedy it. It is part of managing your project, and there are tools and methods to deal with it. According to ISO-31000:2009 (Risk Management — Principles and Guidelines). • Risk is the uncertainty on objectives. Consequently, one can conclude the risk is the probability that something can go wrong with your project, which may interfere or negate the achievement of your business objectives.
10. Resolution	• The ability a person or an instrument to distinguish differences. The capability to discern or distinguish detail.
11. Uncertainty	• The possibility of more than one outcome.
12. Variability	• Something that is unpredictable, able to change or not constant; for example, the budget for your project is reduced due to economic conditions.

Why Risk Management

Glossary of General Risk Management Terms

Although this is a useful glossary of general risk management terms, this is not an FMEA glossary.

Term	Description
1. Risk Planning	• The decision-making process to define the approach (strategy) and plan the risk management activities for your project.
2. Risk Identification	• The process to determine and document which risks will likely affect your project. Also, updating your risk management plan with new risks and observing the progress of your development project.
3. Risk Prioritization	• The process to assign levels of importance to the various risks that are possible in your project, based on clearly understood and written criteria.
4. Risk Management Planning	• The process to develop the reduction of threats to your project by means of procedures, techniques, roles and responsibilities.
5. Risk Monitoring	• The process to detect early signs of risk through the life of your project.
6. Risk Control	• The process to execute the risk reduction plans to mitigate or eliminate the damage of risk to your project.
7. Risk Mitigation	• The steps you take to prevent, correct or deal with risk. Generally, risk mitigation made up of two parts: First, those steps you take to avoid the risk from ever taking place; and second, those steps you take if the risk does happen.

See Also
Glossary

Critical Success Factors (CSFs)

> **Definition — Risk**
> Something that can harm your program and should not happen.

> **Definition — Critical Success Factor (CSF)**
> Something that is critical to your program and must happen.

Many of us forget that risk has a converse: Critical Success Factors (CSFs). While risks are things you do not wish to happen, Critical Success Factors are those things that must happen in order to achieve the desired results. CSFs are the elements within your project crucial for its success.

The intent of creating and managing a list of CSFs is to focus your organization or team on those critical items. There are many distractions during a development project, and it is easy for your team to lose sight of the important activities; thus, the need for CSFs.

The leverage of CSFs is equally powerful to risks in managing your development program. They must be visible, have support from your managers and have an organization-wide agreement — although, it can also be team-wide agreement.

Important note: The definition of Critical Success Factors varies and has changed over time. Furthermore, the phrase Critical Success Factors is used in strategic planning at the corporate level. For the purposes of this guidebook, we use the phrase in the context of a project to develop a new product. Consequently, we take a more tactical definition of CSFs.

The methodology to develop CSFs is the same as that of risk management — naturally, with the exception that CSFs *must* happen. You can use the same processes, identification, prioritization, logs, protocols and other tools to identify and manage CSFs. And you can identify them as you do risks.

Rules for Critical Success Factors

- We always recommend that your list of CSFs be short (fewer than 10, preferably around five). Remember the criteria defines them as the "vital few" things that must happen. Consequently, you must prioritize them so you can manage them. The methods for prioritization are the same as the one used for risks. Perhaps in a large development program, such as a brand new platform or products for a brand new market, your list may double. But always keep it short and manageable.
- As with risks, you must involve your team in the creation, prioritization and review of CSFs. This creates a common set of references and helps the CSFs become integral to your program.
- To develop useful CSFs you must start with the business objectives of your development project — much as you do with risks.
- Your list of CSFs must be visible to the organization. It must have the active support from your managers, and it must be supported by your team.
- CSFs should be quantifiable. You must be able to measure progress and outcomes (results).
- Monitor your list of CSFs, and take appropriate corrective action as needed.

Examples of Critical Success Factors

- The testing of the GUI by our Usability Group, with at least 12 potential users, must be comprehensive and be concluded on time — no later than October 1, 20XX. This will ensure that the new GUI will be ready for incorporation in the "Alpha" product. Fred S. is the owner and champion of this CSF.
- All 12 Major Account Visits must be successfully concluded by January 27, 20XX. Roger J. is the owner and champion of this CSF.

Why Risk Management

The Uncertainty Continuum of Risk

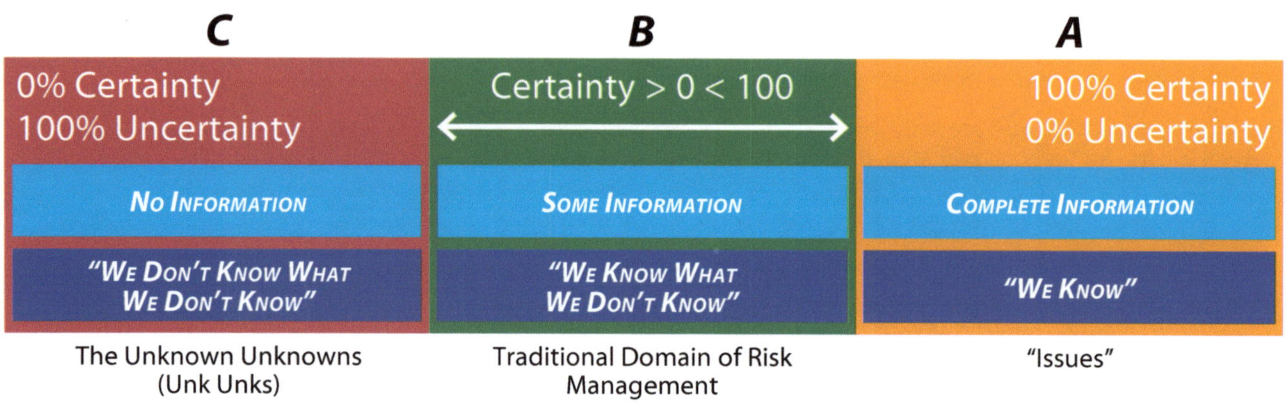

This diagram shows a common depiction of risk as a continuum and is a good way to explain the difference between *risks* and *issues*. It is also a good way to show that there are real risks afoot even if you and your team have not been able to identify them.

Segment A (right column) shows items that have 100 percent certainty of happening. Obviously, these are no longer risks, but they are issues and should be managed separately from your Risk Management Plan.

Segment B shows risks that have less than 100 percent probability of happening and items that have more than zero. This segment is what is generally called "risk management."

Segment C shows risks that exists, you and your team know they exist, but you are not sure in what ways they exist. This area is generally called the "Unk Unks," which stands for Unknown Unknowns.

Managing Unk Unks

Unk Unks (Unknown, Unknowns) is a special category of risk, which is particularly prevalent in technology organizations. The premise for Unk Unks is that there are real risks that could potentially affect your project that you don't even know about yet. These Unk Unks can be divided into two categories:

1. Risks that you and your team are not yet aware of, but could discover.
2. Risks that you and your team will never realize until they occur and immediately become an issue.

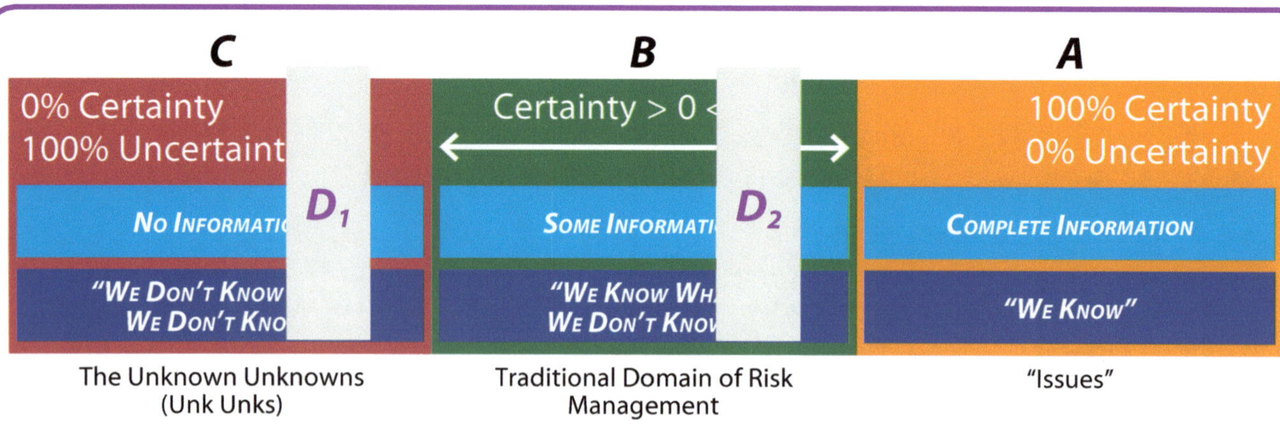

In our Uncertainty Continuum of Risk diagram, the Unk Unks are shown as gray bands in various places in the continuum. In this example, there are unknowns that you will never know (D_1). There are also unknown that you are not aware, but could become aware (D_2).

Why Risk Management

Risks deal with known uncertainties: you know that a vendor's delivery can be delayed or that module test results can force your team to rework the design of a new module. You also have insights that allow you to judge the probability and severity of the risk, which means you are dealing with the controllable uncertainty of known unknowns.

In technology development, the Unk Unks become extremely important, yet the conundrum is: How do you manage risks of which you are not aware? New approaches have emerged in the last few years to address this area. We cannot pretend that risk is not there. Some methodologies in the past simply recommended to disregard this area because it was perceived that there was nothing you could do about it; this is not acceptable, as risks are always there. Always!

Known unknowns can be managed through traditional risk management practices such as preventing risk (stick with proven, but less powerful, module design), transferring risk (penalize the vendor for late delivery) and mitigating risk (add buffer times to the schedule).

Proactive Risk Management and the Unk Unks

Among the emerging strategies to manage the Unk Unks is Proactive Risk Management (PRM) — an aggressive approach to risk monitoring with frequent meetings, frequent updates to the Risk Management Plan and particular focus on the Unk Unks. The principle is that as your project progresses, new information comes to light, some of which will identify the hidden risks in the Unk Unks area. PRM practices such as support structures for continuous risk monitoring and rapid problem solving take risk management one step further and include the management of residual uncertainty. They improve organizations' ability to cope with unforeseeable events that are not reflected in contingency plans such as a vendor's plant going up in flames or the design team losing several of its key players. Though these events are unforeseeable, they are fairly straightforward to deal with; managers understand beforehand that vendor deliveries and design capabilities are critical to project success. They can include these aspects in their risk monitoring and use Proactive Risk Management techniques to react when the risk becomes apparent.

Multiple Paths and the Unk Unks

Another approach is to have multiple paths to a solution. When a new technology is perceived to have high risk, then a second development path is created — some would call this the "Plan B" approach. This is where a second technology or a second approach to the same technology is started, funded and managed until such a time where the first technology has proven feasible.

Unknown unknowns (Unk Unks), on the other hand, are blind spots that nobody is aware of at the outset of the project. They frequently exist in complex projects, where a combination of managed risks and unforeseeable, but generally manageable, events can add up in unexpected ways and are beyond control. They also exist in innovative projects and entrepreneurial activities, where the project objectives are changing on the fly as customer requirements, technological capabilities and business ecosystems evolve.

Additional Methods to Manage Unk Unks

The foundation of Unk Unks management is an advanced, proactive risk management for known unknowns and residual uncertainty, but these practices are not sufficient. Two additional approaches are needed:

1. **SELECTIONISM** characterizes an approach in which multiple alternatives are pursued in parallel, and the one that works best is chosen ex-post. The approach is also characterized as "Darwinian" because alternative design solutions are exposed to the changing business environment, and the fittest solution survives.
2. **LEARNING** characterizes an approach of adjusting the project to the changing environment as it occurs; these adjustments are based on new information obtained during the project and on developing new — that is, not previously planned — solutions during the course of the project.

At first glance, both approaches defy conventional efficiency wisdom, because they seem to create waste (unused alternative designs and time-consuming trial-and-error loops) and reduce the ability of managerial

Why Risk Management

Important

Risk is always there. Your only choice is manage it. Ignoring risk is not an option.

control (no stable project plans and no easy stage-gate planning). They are, however, intended for projects in which the prevalence of Unk Unks causes regular project and risk management approaches that require stable control environments to fail. Consequently, they are not intended for every type of project.

Selectionism and learning are not mutually exclusive. For example, the early planning stages of a new product, the learning curve and the resulting time delay are often relatively more expensive than parallel trials of alternative product concepts, which can take the form of simple concept descriptions that are evaluated by the customer. Manufacturing ramp-up, on the other hand, typically requires high investments and warrants a sequential learning approach. Companies have to decide which approach to Unk Unks management to use for what part of their specific project.

Successful Unk Unks management is based on managers' acceptance of their inability to forecast, plan and control some parts of their project. This requires a change in the mind-set, the infrastructures (reporting systems, incentive structures) and management practices (contracting, problem solving and stakeholder management) of project teams.

Alternatives You Can Use to Manage Risk

You must deal with risk. It is part and parcel of managing projects ,and integral to the reality of your organization. The one thing you cannot do is to pretend it is not real. Your options:

1. **PREVENT IT:** Take steps to ensure that the risk or the impact (consequence) of the risk is avoided. The most desirable option is to ensure that the risk does not happen in the first place.
2. **ADDRESS IT:** Here is where corrective and remedial action come into play. Part of your mitigation plan is to know what you will do if the risk does happen. For example, it might include revising the scope of your project based on the business objectives of your organization in order to lower the risk to a more acceptable level. These are difficult trade-offs and must be made in the context of your business objectives. Sometimes you may have to take the risk in order to achieve your business objectives. Dealing with the risk includes two types of approaches: Corrective Action or Remedial Action. The most desirable step is to eliminate the root cause of the risk (Corrective Action). Methodologies such as the Fishbone Diagram, TRIZ and Problem Solving Methods lend themselves well to root-cause identification. Sometimes we may need to apply a temporary fix, where we treat symptoms of the risk but not the actual root cause (Remedial Action).
3. **TRANSFER IT:** Move the risk to another entity; for example, outside consultants or contractors.
4. **DELAY IT:** Choose to delay the risk by postponing the trigger as much as possible.
5. **SHARE IT:** It may be possible to share the risk with another party; for example, your customer, suppliers, vendors or OEMs may be willing to share the cost of the risk.
6. **ACCEPT IT:** Make it part of your business model and choose to tolerate the impact; that is, "take the bullet." This may be your best option if the expected loss is lower than the other risks or it will be simpler to absorb the impact than to deal with it; in which case you do need to plan for the impact by budgeting contingency reserves.

Pretending that the risk does not exist or ignoring it is not an option. This is the head-in-the-sand approach. Sooner or later, it will impact your project. And ignoring a risk is simply a form of denial, a way for the organization to avoid dealing with reality. The smart project manager ensures that in spite of denial, the risk is defined and mitigated.

Alternatives You Can Use to Manage Risk

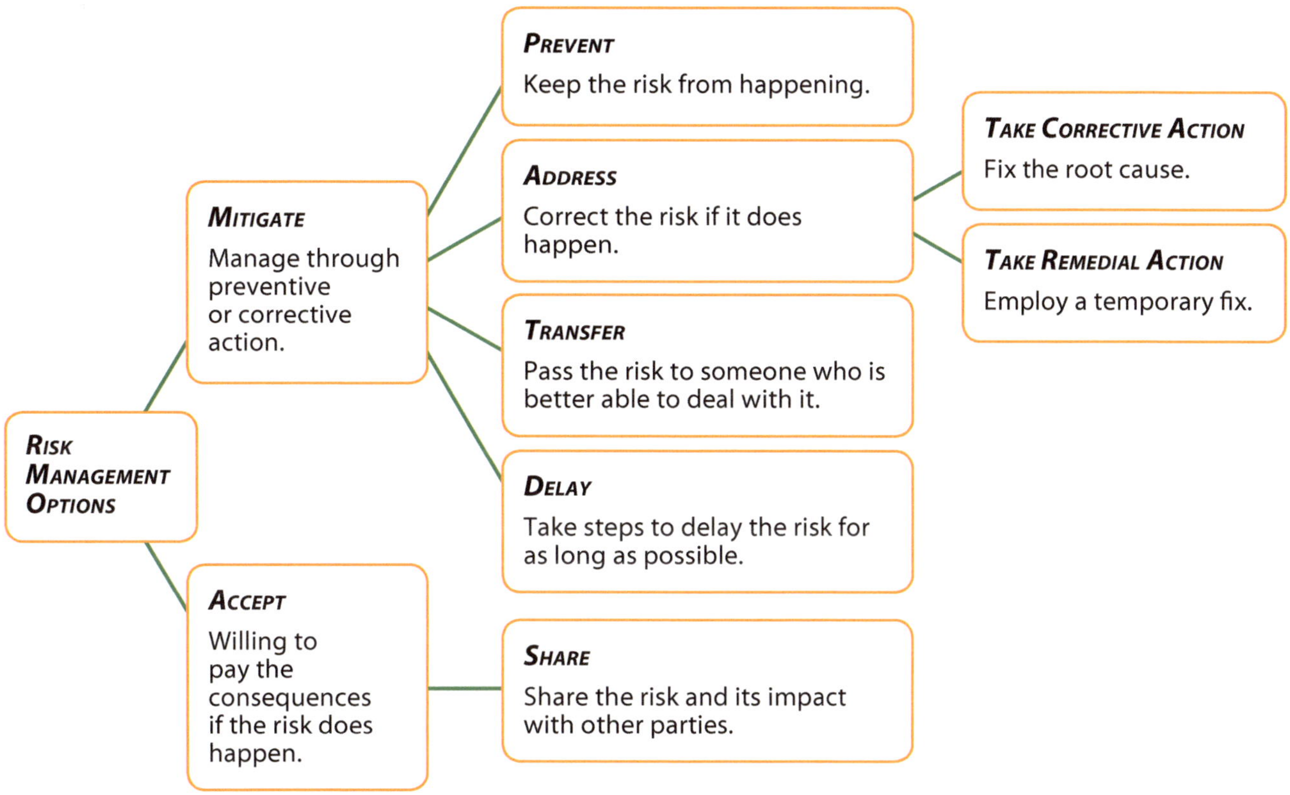

This diagram shows the various options you can choose when managing risk. Your choice is a function of your business objectives and the objectives of your program. Ignoring risk is not a rational option; neither is assuming best-case scenarios and pretending that everything will be perfect.

Why Risk Management

Proactive Risk Management (PRM)

When your project is complex — for example, the technology you are using is not yet stable and the pace of your project is breakneck — you must adopt a Proactive Risk Management (PRM) strategy. It's not enough to develop a risk plan; your team needs to constantly monitor the plan, identify new risks and inform the larger team. In short, PRM is not a passive activity where you develop a plan and then file it away.

Rules for Proactive Risk Management

The first steps in developing a PRM plan are documentation and communication. Your plan must be documented in a way that clearly communicates to your team the actions needed to describe, monitor and manage risks.

Hold regular meetings with your team to monitor your plan and identify new risks. These meetings work best when they are solely for risk management — not part of other meetings. For example, you can have a program review meeting, and immediately following hold your risk management meeting. This will help you focus the team on risk and give risk the stature that it deserves.

Communicate your risk plan consistently, inform your extended team, and keep managers and executives well-informed.

Monitor Your Risk Plan Using the Critical Questions of Proactivity

The list below can also be used as the agenda for your Risk Management Review Meeting.

- What has changed since we last met?
  - What important changes have taken place that impact our project?
  - Are there any new risks that must be added to the Risk Management Plan?
  - Who are the risk owners for the newly identified risks?
  - Are there any old risks that are obsolete and must be deleted from the Risk Management Plan?
  - What changes have happened to the current list of risks?
  - What other changes have take place?
- What are the reasons for the change or variations?
  - What has caused the new risks?
  - What caused the changes to existing risks?
  - What cause the deletion of risks?
  - What are the implications of the changes?
  - Are there any trends that will impact our Risk Management Plan?
  - What will happen if the new trends continue? What should we do right away to mitigate the trends?
- What are the specific changes to our current Risk Management Plan?
  - Do we need to add new risks?
  - Can we remove any risks?
  - Are there changes to current risks?
  - Are there changes in responsibilities?

> **Important:** Your entire organization must have the Risk Management Reflex.

NEW TOPIC

Risk Maturity

At What Level is Your Organization?

Every organization has a level of maturity when it comes to risk management — the ability of the organization to manage risk. There is also the level of competency of the individuals in the organization. These two variables will influence your ability to develop, manage, use and, most important, derive the benefits of risk management.

The diagram below shows an Empirical Maturity Curve that may help you identify your current level and set targets for improvement. For example, many organizations do not have any type of formal risk management model; we show this at the ad hoc level in the diagram. The value of this diagram is to encourage dialog in your organization that will lead to a commitment for improvement. This will help you determined the desired level whereat your organization needs to be to deliver the expected business results. We recommend that you approach the goals in a gradual fashion. Take baby-steps. Set an achievable goal, once you reach it, set another modest goal. Repeat the steps in order to improve them.

If your organization has never done risk management, you can expect a certain level of resistance from individuals and the collective organization. Resistance to change is common. But you must exercise your leadership to ensure that a risk management plan is established. Once people become accustom to the protocol, the task of implementing risk management for future projects will be somewhat easier.

Risk Management Empirical Maturity Model

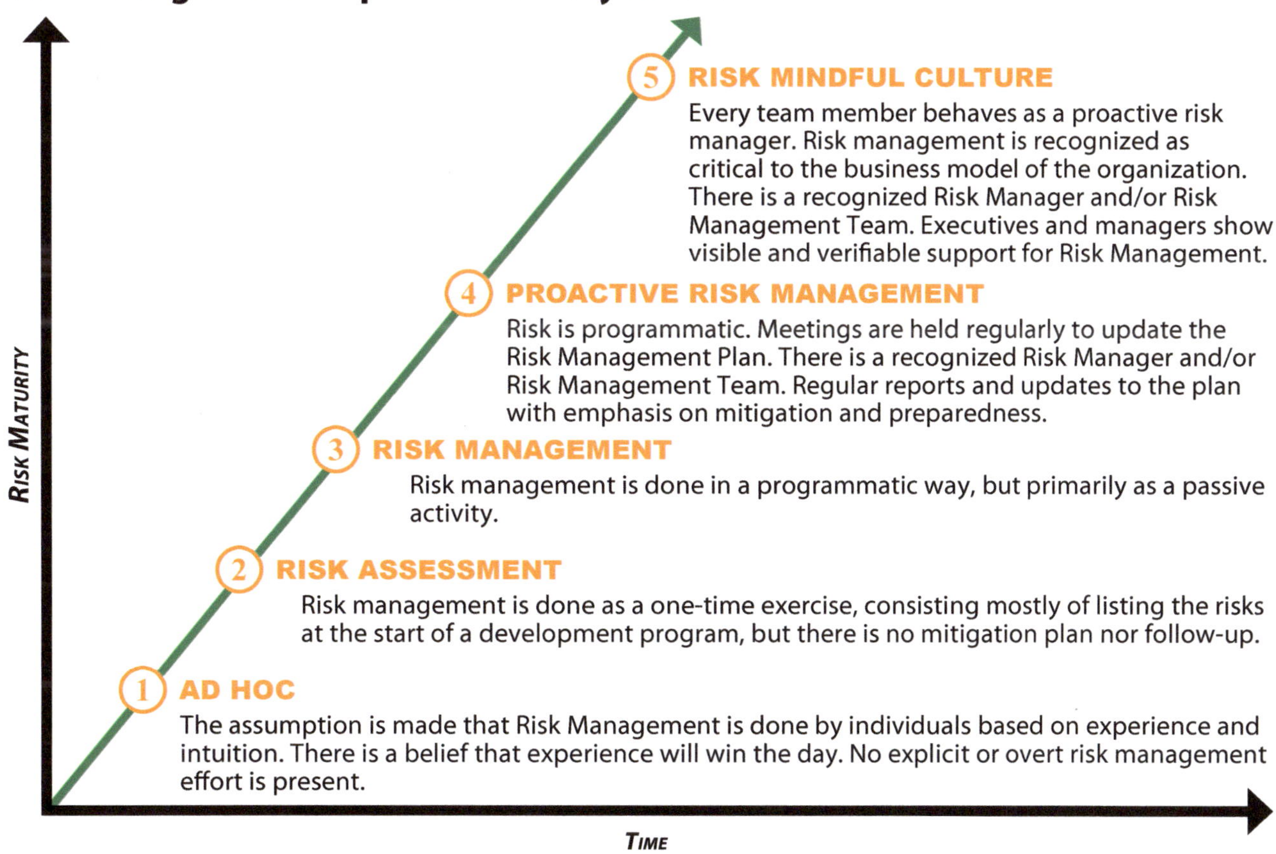

This diagram represents a Maturity Model for a typical product development organization. It was developed from the experiences of the author and will help you determine where your organization lies and, more importantly, where it needs to be to deliver the expected business objectives.

Roadmap to Maturity

Risk Management in Your Organization's Culture (DNA)

Most risk management methods will come up short when the environment is chaotic; i.e., very complex or fast-changing. Things are moving fast enough that even with frequent risk management meetings, you and your team are bypassed by risk. New risks emerge, existing risks change and the Unk Unks rear their ugly heads.

Professions where there is high risk to human life spend considerable time and resources in embedding risk management into their culture (DNA); for example, on the flight deck of an aircraft carrier, where you cannot afford to call a meeting, but rather depend on the culture to anticipate and manage risks as they emerge. In these areas of high risk, you need the entire organization to have the risk management reflex. This instinctive reaction is combined with formal risk management methods. It is not a trade-off — you must do both.

In product development, complexity is driven by several variables — the complexity of your development program, the pace of development, the changing nature of the technology and the constant change of customer requirements, to name a few. This complexity may defeat traditional methods of risk management. It calls for risk management to be part of the culture.

To combat this problem in a product development environment, you need to embed the risk management attitude in the culture or your organization, where everyone is always on the lookout for new risks, always alert to the changing environment. Risk management then becomes part and parcel of managing your project — what noted author Preston Smith calls, "Intrinsic Risk Management."

In short, risk management in your DNA means that it is no longer a novelty, but rather the way you and your team do things; i.e., always sensitized to identifying and managing risks.

How to Create Risk Management in Your Organization's Culture (DNA)

Making risk management part of your culture takes work and time. First, you must make the business decision that it is important and, therefore, worth doing. It is a decision not to be taken lightly, as it will require your commitment, time and tenacity.

Once the decision is made, you should develop a multi-year plan. You are about to change the culture of your organization, and this always takes time. A reasonable time frame may be two years, perhaps three.

The plan must include four critical areas: mission, training, equipment and motivation. Each plays an the essential role in changing your culture. This guidebook is not about changing culture, so we suggest that you avail yourself of a book on the topic. We will provide information on how to get started.

Once you have decided to proceed, we suggest you form a small team to help you in the cultural change, perhaps three to four of you — some people call them "tiger teams" or "core teams." Regardless of the name, it's a small group of individuals who can make the commitment to change the culture.

- **MISSION:** Simply, the articulation of the desired state. Currently your organization has a level of risk management in its DNA — perhaps very low, but a level nonetheless. Your role is to articulate the

Level of Risk Management DNA

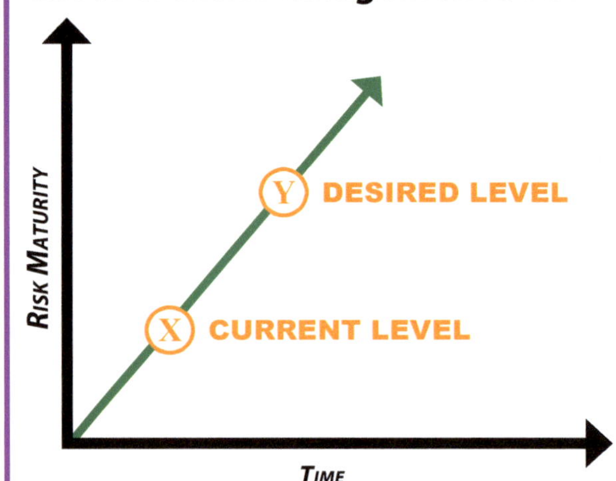

This diagram illustrates the cultural challenge to achieve risk management DNA. It all starts with identifying your current level — the as is. Then, set your target and proceed with the four critical ingredients: mission, training, equipment and motivation.

See Also: Risk Management Empirical Maturity Model

desired level, which is obviously an improvement over the current situation. This sets the target; that is, going from where you are to where you need to be.

- **TRAINING:** Obviously to change a culture, people need to learn the new knowledge. Training may take the form of the traditional classroom setup, and it also includes books, handbooks, coaching and staff meetings. Incidentally, training is not a single event — you need to continuously reinforce the new knowledge. On the other hand, do not fall into the trap of only doing training — this rarely, if ever, succeeds in changing a culture. We suggest that training is only about 20 percent of what you need to change a culture.
- **EQUIPMENT**: The deployment of tools needed to support the new culture — in this case the creation of risk management DNA. For example, you may need to create a new process and a new website with the proper information.
- **MOTIVATION:** A critical ingredient in changing any culture, motivation is the recognition, compensation and ceremonies to continually reinforce the new culture and values. Perhaps you establish a new level of recognition for those who show they understand risk management DNA and act accordingly.

Don't expect your culture to change overnight, nor through a single meeting or training session. It will take time, and your tenacity will win the day.

See Also

Cascading a New Tool

Selection of Risk Methods Using the Complexity Diagram

Not all risk management challenges are the same; thus, it does not make sense to use the same method every time. Every project is different; otherwise, they would be called "processes." And since every project is different, it follows that your risk management approach should also be different, specific to each project.

You must use your judgment in deciding which methodology you will select. Some are very robust and take considerable time; others are simpler and can be done more quickly.

NOTE: You may not have a choice of methods if your company has a standard, your customer demands a specific methodology, or your legal department imposes a standard due to possible legal liability.

The Complexity Diagram[1] might help you make more rational decisions. It is based on a criteria of three — we believe more variables add confusion and little value. You and your team must have a solid conversation on the selection of the methodology. Feel free to use other variables to suit your business objectives and the culture of your organization.

The method of risk management and the overall approach are driven primarily by your business objectives. The selection may be more obvious if other variables are taken into consideration.

About the Scales for the Complexity Diagram

- Scales are qualitative (subjective) and based on the your and your team's opinions and experience.
- Scales are qualitative, but it is possible to develop quantitative scales — although for the majority of projects the qualitative scale will suffice.
- Scales are generally from 1 to 10. It is a very convenient bracket and well-accepted through the business world. While it is not required, we strongly recommend the use of 1 to 10.
- We suggest you start by only defining levels for even numbers. Later, you can add additional resolution and clarity.
- You should define and document each of the levels in your scale. Some prefer to define every other level; for example, defining only the even numbers is acceptable.
- The definition of the scales must be done in collaboration with your team. You should reuse the scales in future projects; naturally you will improve them as you better understand the variables.
- Ensure that your team, managers and executives understand and support the scales. They will be used to make very important decisions, and you do not want misunderstandings.

Examples: The Complexity Diagram

The diagrams shown on the following pages chart four different variables that you will be able to use to determine your risk management strategy. We do assume that every development project is different; consequently, every risk management approach will also be different and will have a different "diamond." The risk management strategy includes the approach you will use to manage risk, such as frequency of meetings, level of detail in your risk management plan (RMP), the resources you should allocate to managing the risks, etc.

You can also use the Complexity Diagram to compare the risk of several development projects. This will allow for better resource allocation, emphasis on risk management, frequency of reviews, etc. Very useful when you are in an organization where there are many development programs at the same time (concurrency).

---

1 The Complexity Diagram used in this guide was inspired by another diagram created by Dr. Aaron Shenhar in the context of "project complexity." We thank Dr. Shenhar and recognize his contributions. Our diagram would not have been possible without his insights and wisdom. Dr. Shenhar is the author of *Reinventing Project Management: The Diamond Approach to Successful Growth and Innovation*.

The Selected Diagram Axes

1. **CRITICALITY OF YOUR PROJECT**, also known as the Strategic Importance of the Program, represents the business and strategic role your project carries. For example, a new product into a new market is more strategic than a simple enhancement to an existing product.
2. **URGENCY (PACE)**, where the faster the pace, the more likely risks will arise and the more important it becomes to manage risks. In a fast-paced project you do not want any risk to interfere with the schedule.
3. **NEWNESS OF THE TECHNOLOGY** to your team, where the newer the technology the more risks it will create. If your team has used the technology in previous products, then it is likely that fewer risks will emerge. "Newness" might be applied to the components of your design or the technology of your design. Some teams apply "newness" to the customer — this is acceptable if it helps you better manage risks that may impact your customers. The issue is the novelty of the technology, either to your team, your company or your industry.
4. **DIFFICULTY OF IMPLEMENTATION** to represent the complexity of developing the new product. For example, a new product where four of five different vendors will contribute is more risky than a new product where only your development team will participate.

Other Variables Worth Considering

- The need for accuracy.
- The cost to design the device or system (risk to your investment).
- The purchase cost of your product (risk to your customer's investment).
- Difficulty to implement (the challenge to make it happen).
- Safety requirements (industry standards, your customer's standards and your ethical standards).
- Complexity of the program or project.

Example 1: Complexity Diagram

The examples in the following pages use three to four variables (axes).

You and your team can use these variables or select others that may be more relevant to your business objectives.

We suggest that you do not change your variables in the middle of a project as it will create confusion and lack of clarity.

The development of your Complexity Diagram should be done in collaboration with your team to ensure alignment.

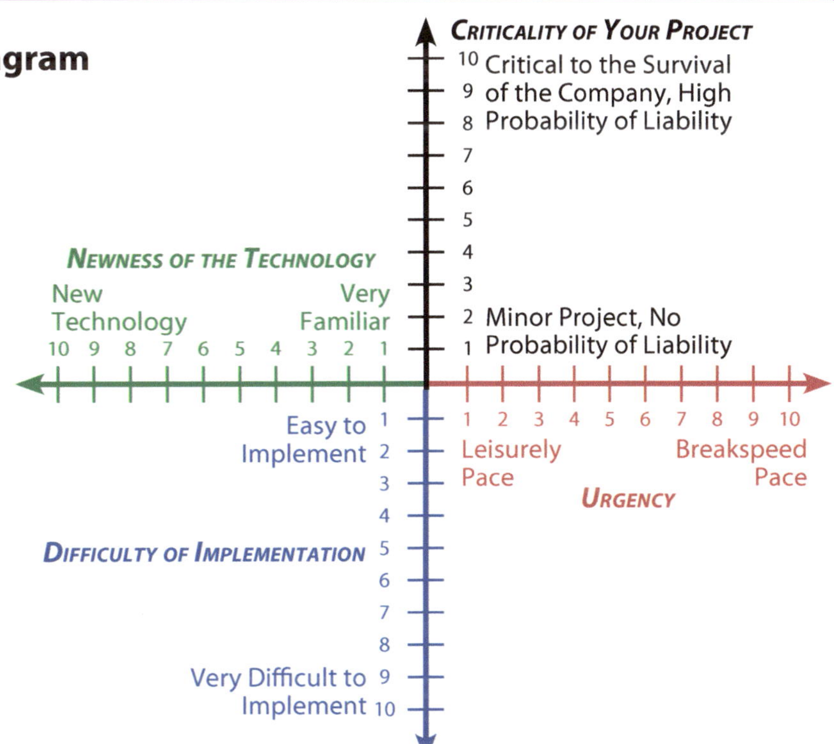

Roadmap to Maturity

Example 2: Complexity Diagram

This example uses only three variables (axes). The urgency is (2), indicating a reasonable pace and ample time. The criticality of the project is (9), indicating very important to your company. The newness of the technology is (10), indicating the technology is extremely new to your development team.

The logical conclusion is to emphasize risk management on the technology at the Program level due to its importance to your company. We would also suggest a robust risk management plan for technology, with frequent reviews and proactive risk management.

The development of your Complexity Diagram should be done in collaboration with your team to ensure alignment.

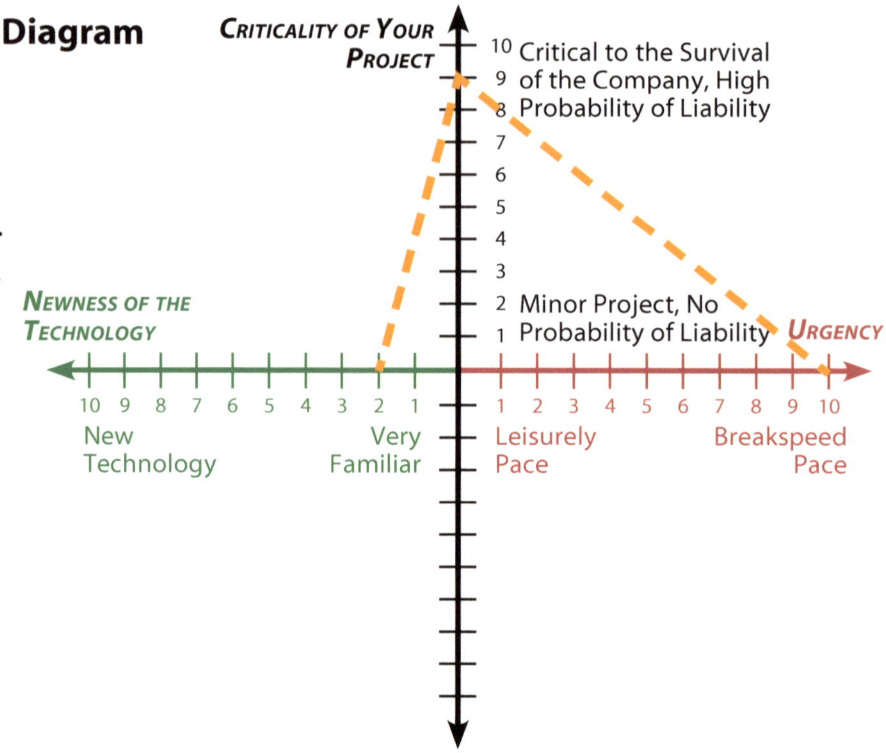

Example 3: Complexity Diagram

This example uses only three variables (axes). The urgency is (10), indicating an extremely high pace and little time to develop the new product. The criticality of the project is (4), indicating reasonably important to your company The newness of the technology is (4), indicating the technology is mostly familiar to your development team.

The logical conclusion is to emphasize risk management on the schedule and things that could interrupt it. Due to the pace, we would suggest emphasis on the technology as well.

The development of your Complexity Diagram should be done in collaboration with your team to ensure alignment.

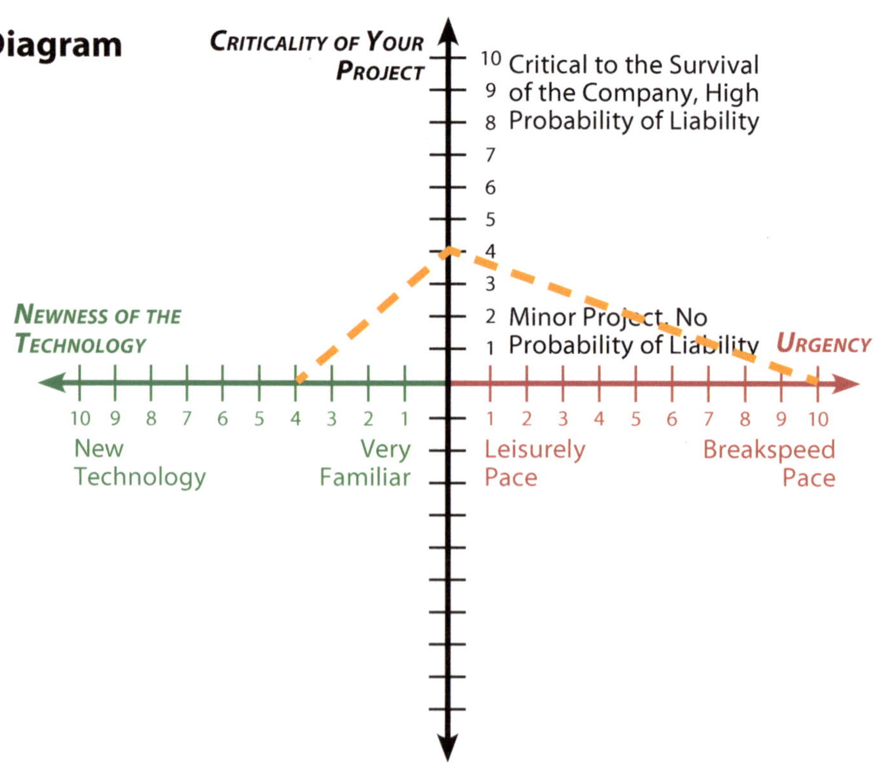

Example 4: Complexity Diagram

This example uses four variables (axes).

The Urgency is extremely high, as is the pace; consequently, our risk management efforts should focus on the schedule and anything that might interfere with it.

The newness of the technology is low, which means we need not devote too many recourses to managing risks in this area.

Although this example uses the scale from 1 to 10, you and your team need to develop your own "qualitative" scale and attach relevant meaning to each level. We suggest that at first, you only define levels for every even number. Later you can add additional resolution and clarity.

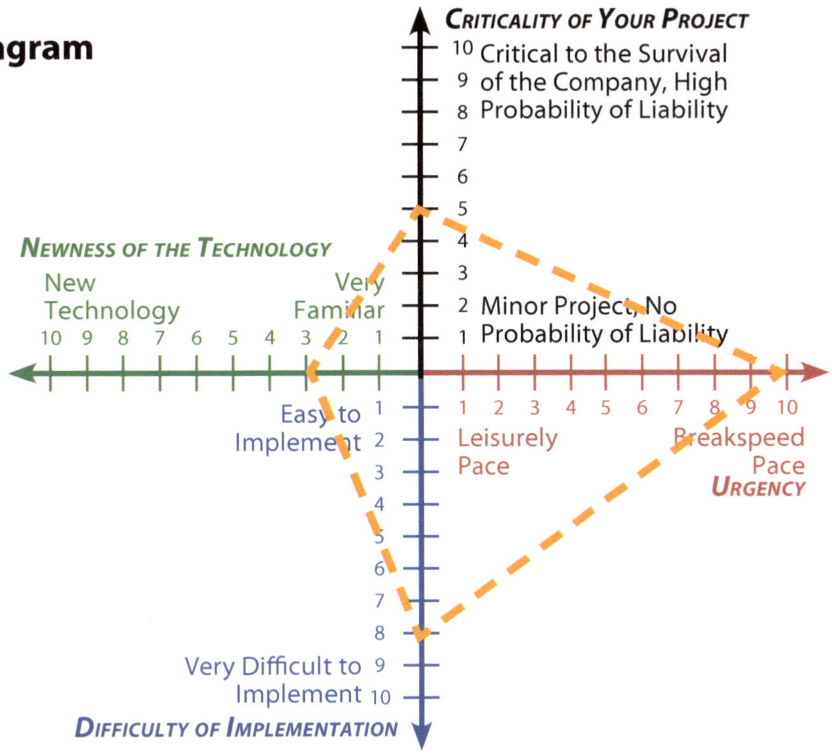

Example 5: Complexity Diagram

This example uses four variables (axes). It represents a very challenging project because two factors rank extremely high.

The team will have to make a comprehensive effort in risk management, primarily because the project has a high strategic importance to the company, and secondarily because the pace is very fast.

This example calls for extraordinary risk management efforts. We would suggest at least chartering a Risk Management Team and securing a Risk Management Mentor from the Executive ranks of the company. Perhaps appointing a full-time Risk Manager or hiring an outside consultant to play the role would also be prudent.

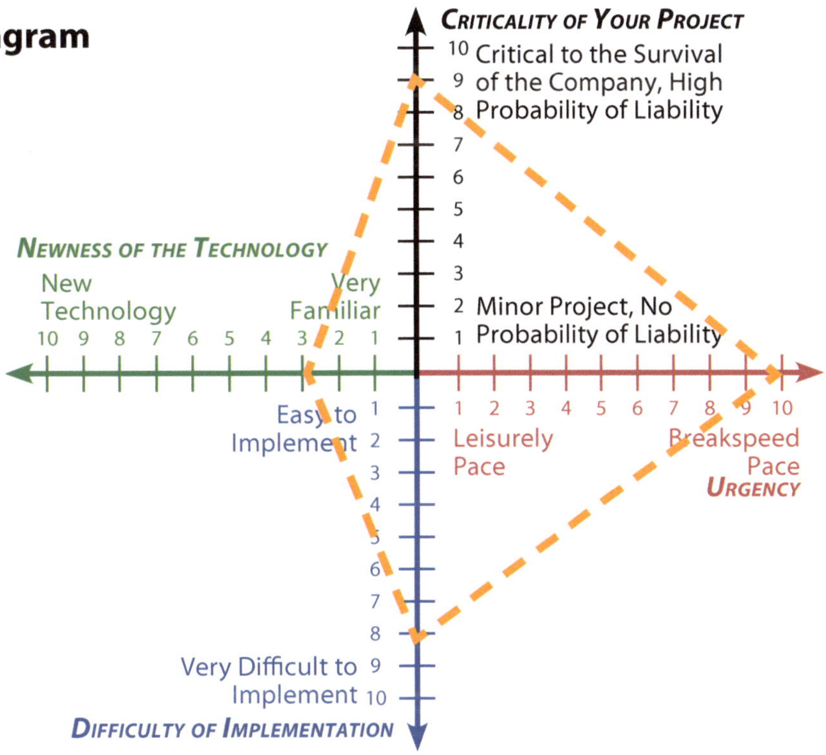

Types of Risk Management to be Performed

Risk management can be performed at various levels, depending on your objectives. For example, you could do a Risk Management Plan at the marketing level that would address the risks of introducing a new product or the risks with entering a new market. Or you could create a risk plan with emphasis on your supply chain, starting with the focus on a component-level risk plan and working toward to a full system-level plan.

There can be several Risk Management Plans associated with a single program. For example, marketing, engineering and manufacturing can work on three plans concurrently. In this situation, it is necessary to provide coordination and cross-functional support to avoid duplication and take advantage of one another's findings.

After you have analyzed your project using the Complexity Diagrams in the previous section, you can then decide on the framework of how you will manage risk. Use our template (on the next page) at the start of a development project to define your overall approach. Complete the template in full collaboration with your team.

The template and example will help you decide the type and level of Risk Management Plan you need to perform. The type, such as manufacturing, identifies the functional focus of the risk management. The level, such as system, identifies the part of the program on which you will focus your risk management resources. Whatever you choose, it is imperative that your team has full clarity as to the type and level being applied.

There may be other types of risks in your line of business. Work with your team to adapt this template to your business objectives and corporate culture.

Example: Types of Risk Management

Types	Level of the Risk Management Plan: Program Level	System Level	Subsystem Level (Module)	Component Level
Financial (Business case)	Product management will develop, document and present for approval a Risk Management Plan on the financial risks of the new product, vis a vis the assumptions in the Product Proposal.			
Marketing (Market)	Marketing will not be required to develop a plan for this program.			
Design (Software, hardware)				Engineering will develop, document and present for approval a RMP for all new components.
Manufacturing (Process)		Manufacturing will develop, document, and present for approval a Risk Management Plan for the new assembly process and for outgoing inspection.		
Supply Chain (Suppliers)			Supply chain will develop a Risk Management Plan regarding all our suppliers in the Far East.	
Service (Service and repairs)			Service will develop, document and present for approval a RMP for depot-level service.	

In this hypothetical example, there is a combination of Risk Management Plans due to the complexity of the program. Notice that there are both technical and non-technical risks to this program; consequently, the need to address other areas is very important. Notice also that the template includes several functions besides R&D; moreover, the responsibilities are distributed among the functions.

Teams and Team Charters

Introduction: Risk Management Teams

Participants in Risk Planning

Later in this document we provide several detailed risk management processes, with adequate details to enable the selection of participants. Generally, you must rely on people who have knowledge about your project. It is critical that you think "diversity." Having people all from a single function — for example, engineering — will tend to create blind spots (areas outside of the engineering domain) that are real risks. Thus, the first rule of participation is to create a group made up of diverse people, functions, perspectives and experiences.

We highly recommend thinking outside the box; for example, you might want to invite suppliers, vendors and consultants associated with your project. Also consider experts such as university professors. At the least, bring all the functions that are associated with your project; for example, engineering, services, repair, marketing, sales and so forth.

The Framework for Robust Risk Management in Product Development

New Product Development Team

All the individuals who have direct or indirect participation in the development of the new product; for example, engineering, marketing, service, documentation, etc.

These teams vary from a few to as many as several hundred people.

1 **Risk Management Team Sponsor (Mentor)**

A manager or an executive-level person who has the time and commitment to coach and help the risk management team successfully develop and implement the Risk Management Plan.

2 **Risk Manager**

The one person who is responsible and accountable for the development, implementation and proactive monitoring of the Risk Management Plan.

In some companies this is a full-time position; others rely on the engineering or reliability departments.

3 **Risk Management Team**

A small group of individuals (three to five) who collaborate with the risk manager in the development, implementation and proactive monitoring of the Risk Management Plan.

The make-up of this team varies depending on the project and the business objectives. Generally, engineering is well-represented.

4 **Risk Management Participants**

All the individuals inside and outside your organization who contribute to or play a role in the development and implementation of your Risk Management Plan.

Participants will come primarily from the new product development team, plus contributors from other areas.

The above diagram details all the major components needed for a risk management framework. This structure is needed due to the complexity of new products, the speed of development and the global nature of new product development. An even stronger structure is needed when you outsource technology, manufacturing or other components of your new product — vendors, suppliers and consultants add to the complexity.

Teams and Team Charters

1. The Risk Management Team

The risk management team is a small number of people (three to five) who are chartered to lead the risk management process. This team owns the risk management charter document and meets as often as needed to ensure the integrity of risk management. We highly recommend that this team holds some meetings on their own, not as part of other agendas.

The three to five members of your risk management team commit to serve for the duration of your project. Be certain to have clarity on the amount of time and the number of meetings that these members will attend; i.e., how much time they will invest in the team.

Do note that when defining (identifying) risks you need wide participation in order to have a complete view of your project or product. In short, the risk management team is the steering group, and participation is the way people contribute to risk planning.

2. The Risk Manager

The risk manager is the one person who is responsible and accountable for all risk management efforts. Generally, the risk manager is the leader of the risk management team. The risk manager is a full-time position in many companies. It may be someone from the reliability group or engineering.

The time spent by the risk manager is a function of the complexity of your project. For example, the more uncertainty, the more time the risk manager will invest. Use your judgment and be able to flex. Another variable is the number of projects that must be monitored for risks; that is, the more projects, the more time the risk manager must spend. In some organizations the risk manager is also the project manager; however, we do not recommend this approach unless the project is very small. You want someone who has the time and skills to deliver the value required by your project.

3. The Risk Management Team Participants

The participants are all those who participate in the planning, monitoring and other functions related to risk management. Most of the participants will come from your new product development team, plus some from other areas as needed; for example, service, reliability, quality and so forth. Also, participants will vary depending on your needs.

Participants will receive action items and other tasks from the Risk Management Team. This is the norm and there should be no confusion.

4. The Risk Management Team Sponsor

The sponsor, or mentor, is an executive-level person who has the time and commitment to help the risk management team successfully carry out its responsibilities. Be certain to negotiate in advance and provide clarity as to the role, time to be spent and expectations. You might want to document the sponsorship in the form of a memo in order to add clarity to communication and make the commitment more visible to all.

The role of the sponsor is to understand the intent of your risk plan, to provide advice, remove obstacles, obtain resources and, in short, do everything possible to ensure success.

Teams and Team Charters

The Skill Set

There are a number of ancillary skills that must be present in your risk management framework to make it work effectively. Just knowing risk management is not enough. One could say that risk management opens the door to address specific challenges.

Time

While the methodologies for risk management are remarkably simple, the skills and abilities that you and your team must have are many. Naturally, the obvious one is time. It is difficult to do good risk management if the bandwidth of your team is very limited. This is particularly noticeable when the team does not, as a matter of course, do risk management, and suddenly it becomes part of your project management framework. We recommend that you discuss the need for allotted time with your team and their respective supervisors and managers.

The investment of time doing risk management will have a sizeable return later on in your project. One of the key benefits of risk management is the reduction of time-to-market due to the removal of unexpected obstacles. Your ability to manage risk will deliver more time for development and less stress.

Problem Solving Methods

As the name implies, there are several methodologies available to define, address and solve a problem. Many books on quality and teamwork cover this tool. Ensure that your team is trained on problem solving.

Problem solving is important for risk mitigation — the course of action that you choose once you have prioritized the risk. Further, problem-solving is used when the time comes to do Root Cause Analysis. Once the root cause has been identified, you have a problem that needs to be solved.

Root Cause Analysis

Most of the time a problem appears in the form of symptoms. Further, most of the time we state a risk in the form of a symptom. Sooner or later, you will need to find the root cause. There are several methodologies available. The most common method is the Fishbone Diagram (Ishikawa Diagram), a simple yet powerful way to identify root causes. Other methods are also available such as Affinity Diagrams, TRIZ, Design of Experiments, etc.

Affinity Diagrams

This is a multi-use team tool that is invaluable. An Affinity Diagram is a decision-by-consensus tool that is user-friendly yet packs considerable force. This collaboration tool gained prominence during the TQM days and continues to serve many teams.

Brainstorming

Teams must understand the fundamental rules of brainstorming. Risk management relies heavily on the ability of a team to collaborate and brainstorm to address risk, and to confront risk when it happens. There are several methodologies available to enable a team to brainstorm effectively.

See Also: **Critical Tools**

See Also: **Problem Solving**

See Also: **Fishbone Diagrams** *Root Cause Analysis*

See Also: **Affinity Diagrams**

See Also: **Recommended Books**

Conclusion: The Tool Set

Risk management is always a team effort; thus, a critical assumption is that you and your organization visibly support the team. It is also assumed that you continually provide training in the area of teamwork and collaboration, as they are essential qualities to good risk management.

Most risk management is qualitative in nature, which requires a team to collaborate in a constructive fashion, not engage in needless debate or, worse yet, conflict.

In addition to the Critical Tools section, this guidebook contains an appendix with recommended books about the major tools outlined above.

Why Risk Management Efforts Fail

No one sets about to fail at doing risk management, yet more often than not teams fail to deliver the benefits of risk management to their projects. This not only has negative impact on the project, but it also creates a negative image of risk management as a methodology that does not work.

#1 Reason for Failure: Not Making Risk Management a True Priority

This is the failure to make risk management an essential step in managing your development program. In other words, it is paying lip service to risk management but not allocating enough resources, effort or importance.

What to Do About It

- ❑ Avoid doing risk management until there is a clear understanding in your organization of its importance and its relation to reducing the overall probability of project failure. Take the time to make the business case; i.e., sell the benefits of risk management. Work with your team to clarify any misunderstandings that may exist about risk management.
- ❑ Ask your managers and executives to openly support risk management. They need to endorse its use and also provide the training and other resources to make it work.

#2 Reason for Failure: Underestimating the Amount of Effort Required

It takes effort to perform proactive risk management. Many teams collect the risks, but fail to develop a risk management plan and meet regularly to update the plan. Another source of trouble is the lack of follow-up to complete all the action items generated by your plan and the regular meetings.

What to Do About It

- ❑ Train your team on the entire process of risk management and ensure that they have more responsibilities than simply generating a list of risks.
- ❑ Clarify the roles of everyone in your team concerning risk management. Ensure that every member of your team knows his or her role in making risk management work.
- ❑ As the risk manager, you need to have the tenacity to ensure completion of all action items. In chaotic environments, your team members have many obligations; consequently, unless you maintain the visibility of action items, team mentors will focus on other areas.
- ❑ Practice proactive risk management by having regular meetings, communicating the success of your risk management efforts and talking to the members of your team about their responsibility to support risk management.

Teams and Team Charters

#3 Reason for Failure: Missing Framework of Support

Risk management will not work without the infrastructure to support a robust effort. For risk management to work there must be a framework of support.

What to Do About It

- ❑ Prior to starting any risk management efforts, ensure that there is a structure in place to support the endeavor, such as the one shown in the diagram Framework for Robust Risk Management.
- ❑ Ensure that your team members have the time to participate in risk management. They need to understand that risk management is part of product development, and cannot be ignored.

#4 Reason for Failure: Past Failures and Waste of Time

Your team members may be reluctant to participate in a risk management effort because they have seen it fail in the past. This may be due to lack of support, poorly managed risk management process or even incompetence.

What to Do About It

- ❑ Demonstrate your commitment and ask your team members for their support. Ensure that they understand that the success of the risk management effort rests with them, not just one person.
- ❑ Show them how this effort is different from previous efforts because you are using a stronger framework, new tools and new processes.
- ❑ Secure a Management Sponsor (Mentor) who is willing to spend the time and effort to help you lead risk management to success for your project. Double check with the Management Sponsor is indeed willing to spend the time and effort to make it work.

The Risk Management Office (RMO)

Risk Management has, not surprisingly, been gaining stature and importance in the world of product development. Due to the types of products or technology they develop, some companies have established the Risk Management Office (RMO). We believe that an RMO is not for every company. It is a business decision, and the proper analysis must be done to justify the investment.

The need for the RMO may also be driven by the risk maturity of your organization, when you need to increase the maturity to a higher level.

See Also
Framework for Robust Risk Management

Types of RMOs

- In some organizations the PMO (Project Management Office) has a Risk Manager who is accountable for the RMO activities.
- In some cases it may be appropriate to set a temporary RMO to ensure the adoption of risk management methods. The duration may vary from one year to up to three, depending on your objectives and the speed of adoption.
- Another choice is to set up a virtual RMO made up a two or three functional managers; for example, the Project Management Officer, the Manager of R&D and the Manufacturing Manager. Other combinations are possible, but again it is driven by your business objectives.

Mission of the RMO

The mission is clear: to manage uncertainty in order to ensure the achievement of the business objectives of your organization.

The Role of the RMO

Generally the role of an RMO include:

- The effective adoption of risk management methods and standards.
- Serve as the organization that determines the processes, tools and templates to be used in risk management.
- Work closely with the PMO (Project Management Office) to guarantee a complementary toolset.
- Measure the value of risk management.
- Ensure that Executives and Managers understand their role and responsibilities within the risk management framework.
- Create goals and objectives to improve the efficacy of risk management.
- Serve as the focal point for Best Practices in Risk Management, which is a framework to obtain, prioritize and disseminate Best Practices across the organization and persons new to the organization.
- Track all quantitative data to add more accuracy to the scales.
- Keep all quantitative data to shift from qualitative risk management to quantitative.

Teams and Team Charters

Team Charter

As with any team, the risk management team must be chartered to ensure clarity between management and the team, and also among team members. Chartering also ensures that no false expectations are created.

Not every development program needs a full charter; use your judgment. Work with your team and managers to determine the strategic importance of your program and the complexity of the development. Reserve the charter process for those that will require considerable resources, but most importantly for those projects that are of extreme importance to your company.

Key Questions for Chartering a Risk Management Team

- What is the specific device or product that is the target of the Risk Management Plan (RMP)?
- What are the deadlines and major milestones?
- Who is the person leading the RMP? Who will be the risk manager?
- Who is the person accountable for results?
- Who is the management sponsor or mentor?
- What is the budget for the RMP?
- Can the team hire outside experts, consultants or contractors?
- What are the boundaries for the team? What is it that the team can or cannot do?
- What will be done with the results of the RMP? This is not a trivial question; before you start any risk management, there should be clarity about the disposition of the results. Who will receive them? In what format? And how will the results be utilized to improve reliability? In our experience, we have found that valuable information is lost due to the lack of planning about disposition.
- Who will maintain records and how? How about archiving?
- What if the team needs to escalate an issue, problem or obstacle; who will receive and act on the escalation?

Charter for a Risk Management Team

Example: Charter for the Alpha Risk Management Plan (p. 1)

This two-page example describes a team charter for an actual risk management team that has been modified to protect confidentiality. When creating a charter for your team consider your business objectives, the complexity of the project and other variables.

The Business Case

The need for a robust risk management framework for Alpha is based on the complexity of the project, plus the tight timeline to deliver the first Beta unit.

The Case for Complexity

1. The Alpha project is on a tight timeline. We are to deliver the first Beta unit to CUSTOMER "A" on September 28, 2007. This is approximately one year faster than previous projects with similar scopes.
2. This project is the first truly multi-vendor project for Super Company, where a system integrator (Acme) will do much of the management. While vendors have been retained for previous projects, Alpha is one where a significant portion of the hardware design and manufacturing are being outsourced (approximately 80%). This includes a contract manufacturer new to SUPER COMPANY, which is located in New York, a sub-supplier in Massachusetts (Acme 2), and one more in Italy.
3. Approximately 40 persons from Super Company are involved in Alpha, plus approximately another 20 located with the vendors.

The Major Contractors

Acme Technologies; New York (System integrator)

Acme 2; Massachusetts (Acoustic enclosure)

Acme 3; Italy (Power supplies)

Acme 4; New Hampshire (Turret)

Objective for Risk Management

To develop, document and execute a proactive Risk Management Plan for the Alpha project, which will measurably help achieve the September 28, 2007, delivery to CUSTOMER "A," while achieving the targeted ease of use and reliability as documented in the MRD under MTBF and MTTR.

Overall Risk Management Strategy

1. The scope of the proactive Risk Management Plan includes a design phase FMEA, a program-level Risk Management Plan, and a possible software FMEA.
2. It also includes sustained reviews and risk mitigation for the life of the project.
3. It will be a facilitated process, which relies on a Core Team and an experienced facilitator.
4. SUPER COMPANY will appoint a risk manager within the ranks of SUPER COMPANY.
5. It uses a customized risk management approach that relies on FMEA practices.
6. We will establish a proactive risk management strategy of continually monitoring the Risk Management Plan and defining preventive and corrective action, plus engagement of the Core Team, the risk manager, and the team at large.
7. Training will be provided, but only when needed.
8. Suppliers will be engaged and involved in the process.
9. Suppliers are expected to develop and manage the risks associated with their part of Alpha.

Budget

Budget for risk management will be derived from the overall project funding. The project manager will ensure that risk management is properly funded.

Teams and Team Charters

Example: Charter for the Alpha Risk Management Plan (p. 2)

Critical Success Factors

Management Sponsor	Bob S.
Overall Program Manager	Jane G.
Product Quality Assurance	Sam J., Engineering Mgr.
Core Team Leader	Andy W., Mechanical Mgr.
Risk Management Team	Fred T. Amy P. Mark F. Cathy C.

Overall Facilitator

José Campos

Critical Success Factors

Risk management must be included as a distinct line item (work package) in the work breakdown structure (WBS).

The Risk Management Team will meet every other week, separate but in full collaboration with the project team.

There will be a full-time risk manager for this project for the duration of the project.

Decision-Making

Decisions about risk management will be made by the Core Team; the leader of the Core Team is Andy W.

Key Assumptions

All vendors will be asked to develop a Risk Management Plan, which will be reviewed by the Risk Management Team.

Key Constraints

Risk Management

Overview of Risk Management 50
Identifying Risks 58
Prioritizing Risks 65
Planning Preventive and Corrective Action 67
Implementing Preventive Action 72
Monitoring the Risks 73
Taking Remedial or Corrective Action 75

Overview of Risk Managament

NEW TOPIC

Risk Management

Overview

The process to develop and implement a Risk Management Plan can be reduced to three phases, as shown in the graphic to the right. This view provides a simple, yet realistic view of risk management. Additionally, we use this graphic to help you navigate through the phases as they are outlined in this guidebook.

Note that some suggest that there are only two phases, followed by implementation of the plan. We acknowledge that this is also valid, and the end is the same; that is, managed risks.

Phase 1: Assess

Identify and prioritize the risks. This is where your team has the most participation and you lay the foundation for all subsequent steps. Most risk management methods rely on a **QUALITATIVE** approach; that is, they rely on the experience and collaboration of your team to identify risks. **QUANTITATIVE** risk management falls under the category of "reliability engineering," which is not covered in this guidebook.

Phase 2: Plan

Once the risks have been prioritized, you assign the mitigation for each risk. Mitigation includes preventive and corrective action. In plain English, it answers two key questions: What can we do to prevent this risk from happening? And what are we going to do if the risk does happen?

This phase also includes the documentation of your plan such that it is available to your entire team.

Phase 3: Monitor and Control

This phase is the ongoing effort to monitor risks and update new risks. Here is where the concept of proactive risk management comes into play. It is not enough to develop a plan and file it. You must continually monitor risks and add new risks as they are identified. A schedule of regular meetings is the best way to implement the proactive risk management concept. Wise organizations schedule meetings for the sole purpose of monitoring and updating the risk plan; this way you do not shortchange your ability to manage risk.

Obviously, this phase is where you put your plan into action should any of the risks need mitigation.

What You Will Learn

In this guidebook we provide a fully documented and detailed process of each of the steps in the three phases.

First, you will discover the detailed process map for risk management. Following that, the risk description and risk prioritization are explained. Also see Critical Tools at the end of this guidebook for additional information.

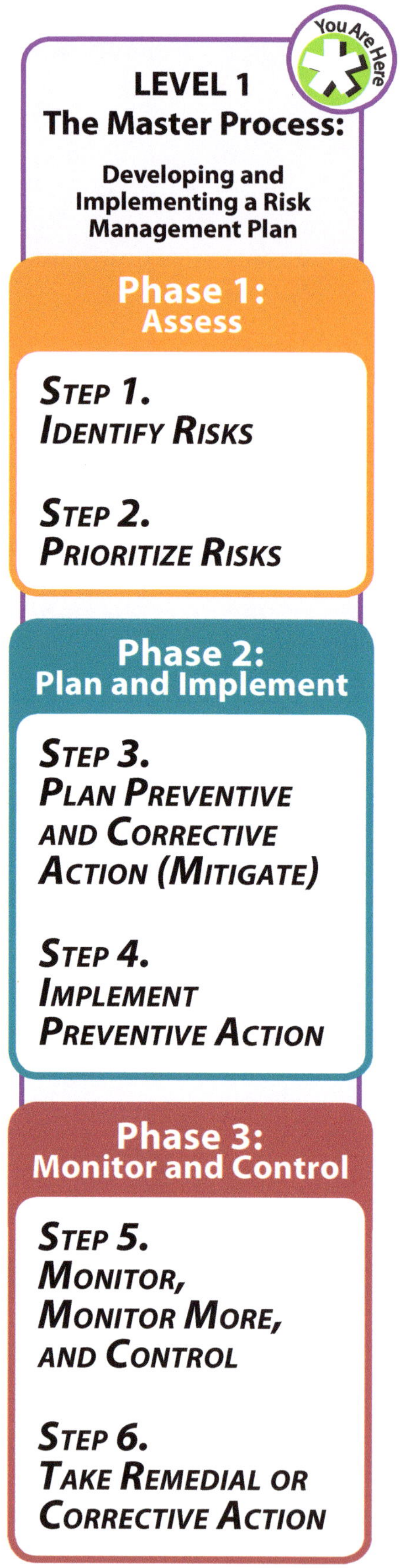

Overview of Risk Managament

Before You Start

Checklist

- ❑ **CALL A MEETING:** Your objective is to call a meeting with the participants in your risk management efforts. This meeting is the first tangible manifestation of your efforts; consequently, you must prepare to ensure a successful event.
  - This first meeting is also a brainstorming session. You want to invite stakeholders plus your risk management team (if anyone has been selected). The intent of this meeting is to get as much input as possible from as many stakeholders as possible.
  - Also note that this first meeting is bound to be long, from four hours to as many as eight. Your guests should be aware and agree to be at the meeting for the duration. It could also be two sessions of four hours each for additional information).
- ❑ **COORDINATE LOGISTICS:** For example, schedule a venue that is large enough for your team to work, and make sure you have the right supplies. What you will need:
  - A large conference room to allow for much interaction.
  - An ample supply of yellow stickies, as it is the principal tool for the task.
  - One or, preferably, two flip charts with fresh paper pads.
  - Markers.
  - Masking tape.
- ❑ **SELECT ATTENDEES:** Select the names of attendees on your risk management team who will attend the meeting; do this based on their ability to help contribute to risk management planning. Ensure diversity of participants — the more points of view the better. For example, combine people who are involved in your project with people who aren't. People from other functions may be helpful, as would some of your suppliers and vendors. Use your imagination to ensure diversity. Who to invite? For example:
  - The members of your team (Marketing, R&D, Service, Quality, Manufacturing, Etc.).
  - Key stakeholders.
  - Key managers or supervisors.
  - It may make sense to invite key suppliers, contractors and consultants. They play a critical role in the success of your development project.
- ❑ **PREREQUISITE TASKS:** Prior to calling this meeting you should have accomplished the following:
  - Completed the Team Charter Document.
  - Selected your Risk Methods and Strategies using the Complexity Diagram.
  - Completed the Types of Risk Management to be Performed template.
  - Already have a good idea of the members of the Risk Management Team, if you chose to have one.
  - Already have the name and title of the Management Mentor (Sponsor), if you chose to have one.

Overview of Risk Managament

Important Notes

Try to keep the group to fewer than 10; more participants adds a level of complexity that may make it difficult to succeed. If at all possible, retain a facilitator. This will expedite the process and make it more productive. Some companies have internal facilitators; your HR department may have a qualified person. You may wish to ask a manager or another project leader to act as facilitator.

Avoid the creation of a bureaucratic document; rather, the Risk Management Plan is a dynamic tool that helps you monitor and take the appropriate action. We encourage flexibility in the definition of the format of your plan. In this Guide we provide two approaches and several examples to allow you to customize to your and your team's specific needs.

This step-by-step process will guide you through the development and implementation of a risk management plan. It may take several calendar weeks to complete the entire process, and that's acceptable. Also, you may need to make minor changes to the process to adapt it to your business objectives and the culture of your organization.

Once you have made the decision to have a risk management plan, the first step is to make all the preparations for the meeting. This is the first active result of your efforts that is visible to the entire development team. By the time you are ready to call the meeting you may already have organized the risk management team, although it can be done as an outcome of the meeting.

Detailed Process: Risk Management

1

PLANNING AND STAGING

Being well-prepared will make risk planning much more effective. As with any project, planning is critical.

Once planning is complete, proceed to invite the participants to your first risk management meeting.

Important

Part of staging is the completion of the charter document.

Ensure the support of the proper managers and supervisors.

See Also **Teams and Team Charters**

2

INVITE THE PARTICIPANTS TO THE PLANNING MEETING

This is your risk management meeting. Allow enough time for creativity, interaction and dialog. These meetings last from four hours up to a full day. Don't shortchange the process — the time spent in risk planning will pay off later.

Who should attend? Those people who can contribute to identifying anything that could go wrong with your project.

Important

Risk identification is a brainstorming exercise; allow for much interaction and ample time for dialog.

During your brainstorming, review previous projects to help you identify additional risks.

See Also **Type of Risk Management to be Performed**

CONTINUED

Overview of Risk Managament

Cont'd

3

Start the Meeting

Start by stating the purpose of the meeting, which is to identify and prioritize risks for your project.

Present a quick overview of the project; for example, use the Gantt Chart to show the schedule. You can also show a Block Diagram, and any other information that may add clarity of understanding.

1. Present the Team Charter
2. Present the Risk Complexity Diagram
3. Present the Type of Risk Management to be Performed

Ensure that everyone has sufficient understanding of your project so they can better contribute.

Describe the project and its objectives, adding any information that might be helpful; for example, block diagrams, photographs, videos or flow charts.

See Also

Protocol for Risk Definition

Time Frame

The process of identifying possible risks could take two to four hours. Don't rush it! This is the most important step in the process. People must have time to brainstorm.

4

Initiate the Brainstorming to Identify Potential Risks

1. Ask the participants: "What are all the things that could go wrong with this project?"
2. Ask everyone to write their ideas on yellow stickies. It is best to request three ideas per person to start, one idea per yellow sticky. By now they should be familiar with the Yellow Sticky Protocol, but if not go over the protocol before they start writing.
3. After a reasonable period of time, ask each participant one at a time to read aloud his yellow stickies and place them on the board or flip chart.
4. After everyone has read their three ideas, ask for more ideas. Encourage people to write down anything and everything that could go wrong with the project. Every new yellow sticky will create new ideas, and this is part of the process. Continue to encourage your team members to create more risk possibilities. Reassure them that there are no bad ideas, and that the rest of the process will ensure objectivity.

Keep brainstorming until it is obvious that all ideas have been captured.

5

Continue the Brainstorming

At this point you should have many yellow stickies on the board or flip chart, perhaps 50, 100 or more.

Make sure they are all readable and understandable. If you need to change a yellow sticky, you must ask the author for permission and clarification. This is very important, as it may generate new ideas and add much clarity.

Important

Once you've finished brainstorming and all the yellow stickies are on the board, take a coffee break! Let your team rest in preparation for the next step.

Continued

Overview of Risk Managament

CONT'D

6

CLEAN UP YOUR BRAINSTORMING

Start by discarding the obviously unlikely risks; for example, "We get hit by an asteroid." But you must ask the author for permission and clarification — this is very important.

Analyze by:

- Reviewing the description: Is it clear? Does the team understand the risk in the same way?
- Should there be criteria associated with the risk?
- Are there degrees to the risk?

See Also **Karnaugh Maps**

Use the 2x2 Matrix to prioritize 30 or fewer identified risks.

Important

You have two ways to prioritize. For a relative few number of yellow stickies — 30 or fewer — we recommend the 2x2 Matrix for prioritizing results (see Karnaugh Maps).

When you have a larger number of yellow stickies, we recommend the PI Factor approach, where P is the probability and I is the impact. The following steps take the PI Factor approach to prioritizing results.

7

PRIORITIZE BY PROBABILITY

Now prioritize the remaining risks by their likelihood of actually occurring.

You can use a scale from 1 to 10, where 10 is the highest probability. Work with your team to determine the value for each of the yellow stickies. Write the value for the probability in one of the corners of the yellow sticky.

You can do a second round of prioritizing, this one using a more scientific analysis of the priority; for example, analysis based on a percentage, where each level is defined using documented criteria. This step is not necessary every time, but only when additional accuracy is needed.

NOTE: This is one of the two priority-setting steps.

See Also **Developing Scales for Risk Management**

8

PRIORITIZE BY IMPACT

At this point, you have prioritized by likelihood — i.e., the probability of the risk happening.

Next, go over all the prioritized ideas, but this time prioritize by asking, "What will be the damage if it does happen?" That is, how severe will the impact be if it does happen?

As you did in the prior step, assign a number from 1 to 10, where 10 is the most severe impact and write the value for the impact in one of the corners of the yellow sticky

Identify the few that have a high probability of happening and at the same time a high level of damage if they do happen. These are the "vital few."

9

CREATE THE RISK FACTOR

By now, each yellow sticky should have two numbers. One for the Impact and another for the Probability.

To create the Risk Factor simply multiply the two numbers. Write the product on the yellow sticky.

Once you've calculated and recorded the Risk Factor for every yellow sticky, you can enter this data into a spreadsheet.

CONTINUED

Overview of Risk Managament

Cont'd

Where You're At

At this point you have:

- Identified the risks.
- Prioritized them by probability.
- Prioritized them by impact.
- Created a PI Factor for each yellow sticky.

Important: The impact can be measured in several ways — your choice. The most common way is to translate into dollars. The impact of every risk can always be translated into the language of dollars. Another choice is "lost days" or "slippage" — the number of days your project will be delayed if the risk happens. Another choice is the impact on the customer — i.e., the negative impact on customer satisfaction.

10 Document Key Assumptions

In many cases, risks are predicated on something else happening. That is, the risk will occur only if something else takes place. These are called key assumptions. For example, "Our budget gets cut" leads to the assumption that the damage or impacts on your project are predicated on the budget being cut. Another example is "There is a company-wide layoff." In this case, the damage to your project is predicated on the assumption that it will occur "if" there is a layoff.

Document all these assumptions. They will be very important in managing risk as your project progresses.

Important: It may be a good idea to end the meeting at this point. If you detect that your team members are fatigued, it is best to schedule a second meeting to define preventive and corrective actions.

Where You're At

At this point you have prioritized each risk based on Probability and Impact. Each yellow sticky will have three numbers written on it: the Probability, the Impact, and the Risk Factor, which is the product of the Probability and Impact. The Risk Factor provides the priority. Naturally the higher the Risk Factor, the more important the risk is.

11 Start Development of the Risk Management Plan

It's essentially impossible to create a Risk Management Plan for every yellow sticky, particularly when you have over 100, so you and your team need to decide how many of the prioritized yellow stickies you will address as part of the plan.

You have the Risk Factor for each; naturally, those with the highest Risk Factor get the priority. You can use your documentation (Charter, Complexity Diagram, and Types of Risk to be Performed) to help you and your team decide.

Once you've prioritized and decided how many you will address in your plan, it's time to start defining mitigation — your Risk Management Plan. For each of the risks you identified as being high probability and high impact, develop a documented action plan for preventive measures and corrective actions.

Continued

Overview of Risk Managament

Cont'd

12

Define Your Mitigation Action

PREVENTIVE ACTIONS are those activities you will undertake to prevent the risk from happening — i.e., you anticipate and thus prevent.

CORRECTIVE ACTIONS are those activities you will undertake if the risk does take place.

For each of the risks you identified as being high probability and high impact, develop a documented action plan for preventive measures and corrective actions.

Once your plan is complete, review with the project sponsor, your manager and those key stakeholders you deem appropriate.

Distribute copies of the plan to your team members.

Archive a copy in a safe place. In some cases, this is a required step due to regular formal meetings throughout.

Important

The most valuable benefit from risk management is discovering the steps you can take, right now, to ensure that the risk never happens.

Pay particular attention to preventive action — you will have the highest return on you investment by making certain that the risk stops being risk. Ensure, right now, that it will never happen!

Where You're At

At this point you have:

- Defined preventive and corrective action for each prioritized risk.
- Documented key assumptions (via yellow stickies and flip charts).

The meeting ends when the above list has been completed.

13

Document Your Risk Management Plan

You now proceed to document your plan. There are several templates in this guidebook you can use. It's a good idea to solicit the help of a few members of the team to help in the documentation — it serves to create ownership, plus it adds clarity to your document.

The document can be an Excel spreadsheet, provided it's easy for all members of your team to understand.

14

Deploy Preventive Action

Once you have a documented plan of preventive actions, lead your team in implementing these actions. Also implement new preventive actions discovered after the document plan was drafted.

Here is where your risk management efforts start to yield results — you save time and aggravation, and you help achieve your desired results.

Continued

Overview of Risk Managament

Cont'd

15

Monitor and Report

Constantly review the list of prioritized risks and assess the need for preventive action. This is done formally and informally.

Formal review includes regular meetings with the Risk Management Team, the engineering team and other key stakeholders. The formal meetings should happen regularly. Use your judgment to decide the space between meeting; that is, weekly, every four weeks or something else.

Informal review occurs through direct interaction with all stakeholders.

Continually review your plan to derive its benefits. Do this through your regular risk management meetings.

Keep all stakeholders informed about risk.

Change and communicate the key assumptions as your project progresses. Changes to the assumptions also mean updating your plan.

The preventive and corrective actions may need to be changed as your project progresses.

Your agenda for your review meetings:

1. Review your plan.
2. Discuss variances.
3. Mitigate risks as needed.
4. Terminate risks that have been mitigated.
5. Identify and prioritize new risks.
6. Update your plan and create new plans for newly identified risks.
7. Publish your plan.

Important!

Remember the Five Questions of Proactive Project Management

1. What is the variance between the baseline and the actual (schedule, budget, etc.)?
2. What are the issues causing the variance?
3. What is the current trend; that is, the preliminary predicted status if we continue with our current performance?
4. What new risks may pop up in the future, and how could they change the preliminary predicted status?
5. What actions should we take to prevent the predicted state from happening and deliver on the baseline?

16

Take Action

YOUR PLAN IS OF NO USE IF YOU DO NOT ENSURE ACTION — A.K.A. IMPLEMENTATION!

- Take preventive action as soon as you complete the plan.
- Take corrective action when a risk cannot be avoided.
- Continue to hold your risk management meetings to monitor and mitigate risks
- Update your plan as needed.

Learn and Improve (Kaizen)

Your risk management process will improve significantly if you constantly add the lessons you've learned.

- After your project is completed, identify how well your Risk Management Plan worked.
- Interview team members and capture their suggestions for improvement.
- Learn from the experience — document and use lessons learned.
- Build Best Known Practices (BKPs) for risk management so that you and others will have the information available for your next project.

Identifying Risks

NEW TOPIC

Protocol for Risk Definition

Introduction

Every risk management methodology starts with the identification of risks; this is intended to capture everything that might be remotely considered a risk. It is a brainstorming phase that relies on the knowledge of a group of people.

Some organizations, based on experience, have a checklist for risks. These lists are compiled over time and based on similar projects. There is a danger (risk) with this approach: No two projects are the same, but a checklist may create a feeling of completeness; that is, you might mistakenly assume the list you use is perfect and complete. The use of a checklist is acceptable, but keep in mind that new risks appear with every new project.

Part of risk identification may include research, either prior to the meeting or afterward. Research includes reviewing prior projects and risk management plans, and also interviewing people in your organization who participated on previous projects or have a special knowledge about a topic of interest. Read papers and articles on the topics of interest, do a Web search and even look into purchasing books on the topic. You can assign research to various members of your team and ask them to come prepared to make a presentation during your meeting.

Speaking of Risk

It is important that you and your team have clarity about the language you will use in defining risks. First, there is the risk itself, "causes the mechanical design to be delayed by five days." Then there is the "cause" of a risk or, as some call it, the "root cause"; for example, "the chief mechanical engineer calls in sick." Think of this as the source of the risk. Finally, there is the impact or effect of the risk, "resulting in a delay of three days for the entire project."

Thus risk, for the purposes of risk management, must be stated in three parts:

1. **RISK:** The actual risk.
2. **CAUSE:** The cause or root cause or source.
3. **IMPACT:** The consequence or effect.

For example:

1. **RISK:** "Competitor A confuses the market."
2. **CAUSE (SOURCE):** "Competitor A introduces a new technology."
3. **IMPACT (CONSEQUENCE):** "Our sales are reduced by 25 percent in the first six months."

LEVEL 1
The Master Process:
Developing and Implementing a Risk Management Plan

Phase 1: Assess
- STEP 1. IDENTIFY RISKS (You Are Here)
- STEP 2. PRIORITIZE RISKS

Phase 2: Plan and Implement
- STEP 3. PLAN PREVENTIVE AND CORRECTIVE ACTION (MITIGATE)
- STEP 4. IMPLEMENT PREVENTIVE ACTION

Phase 3: Monitor and Control
- STEP 5. MONITOR, MONITOR MORE, AND CONTROL
- STEP 6. TAKE REMEDIAL OR CORRECTIVE ACTION

Identifying Risks

Another example:

1. **DESCRIPTION OF THE RISK:** "The risk that we run out of fuel during our trip." Notice that this part only says what the risk is; it does not go beyond a simple description. Also note that it is short, but easy to understand. We recommend you avoid cryptic descriptions — e.g., "Run out of fuel."
2. **POSSIBLE ROOT CAUSE:** As you state a risk you are also making mental assumptions about the cause of the problem. It is not required that you state the root cause, but it is highly recommended.
3. **IMPACT:** "It causes us to lose $500 due to late fees." This part describes the actual damage or consequence as a result of the risk actually happening.

Every risk must have at least one impact (consequence). A single risk can have several impacts, each calling for different mitigation. Stating the actual impact for every risk ensures completeness; that is, avoid the trap of assumptions, where one person believes that his perspective of a risk has been stated by someone else.

It is important that there is agreement within the team as to how the risks are stated to ensure clarity and convenience. Everyone on your team must be able to easily understand the meaning of the risk as it is described. Remember also that others in your organization and beyond (e.g., vendors and suppliers) may have to deal with the risk. Enforce the three-part description, but accept occasional exceptions.

Using the Yellow Sticky Protocol to Document Risk

The need for a protocol, or set of rules, to describe a risk is necessary to ensure communication among the team members and with others not in the team but with whom you will need to collaborate to deal with the risk. A protocol saves time during the development, as everyone on the team knows what they are reading, and in what order.

The ability to ensure that everyone in a group participates is greatly enhanced through the use of Post-It Notes® (trademark of 3M), known in the vernacular as "yellow stickies." A benefit of the yellow stickies is that they provide a self-documenting process. Also, yellow stickies significantly accelerate the documentation, rather than writing things on a flip chart, which is time-consuming. Finally, yellow stickies are portable; that is, one can move them, hand them out, rewrite them, etc.

The risk management process used in this guidebook is **QUALITATIVE** in nature, which means the actual risks will be identified by your team members during the risk management session. As such, there will be a variety of ways that your team members will use, from cryptic "the pump" to very detailed "the two-stage back up pump will fail if the pressure exceeds 200 PSI." Incidentally, for **QUANTITATIVE** risk one needs to look more to the area of reliability engineering, which is not part of this document.

As we recommend in the process map, all risks will be written in yellow stickies; furthermore, we strongly suggest that you create and enforce a protocol for the way the risks are described. This will ensure clarity and uniformity that will save you time and aggravation.

In the following pages we show examples and a description of a protocol. You need not use the approach we describe, but you need to use a protocol for the sake of clarity and efficiency. It is critical that every yellow sticky has the following information: the description of the risk, the possible cause, the impact if the risk were to happen and the name of the person who wrote the note. Feel free to require additional information if it helps you with clarity and efficiency.

Important

In some organizations this protocol is known as the "if / then" method — if such and such happens, then such and such takes place.

Identifying Risks

Examples: Using Yellow Sticky to Document Risk

Example 1: The Protocol in General

Shown here is the Yellow Sticky Protocol for documenting a risk. Notice the description shown in the various call outs. Feel free to modify your protocol to add clarity and efficiency. The intents are to clearly describe the risk and its impact and to trace the yellow sticky to its source.

Example 1

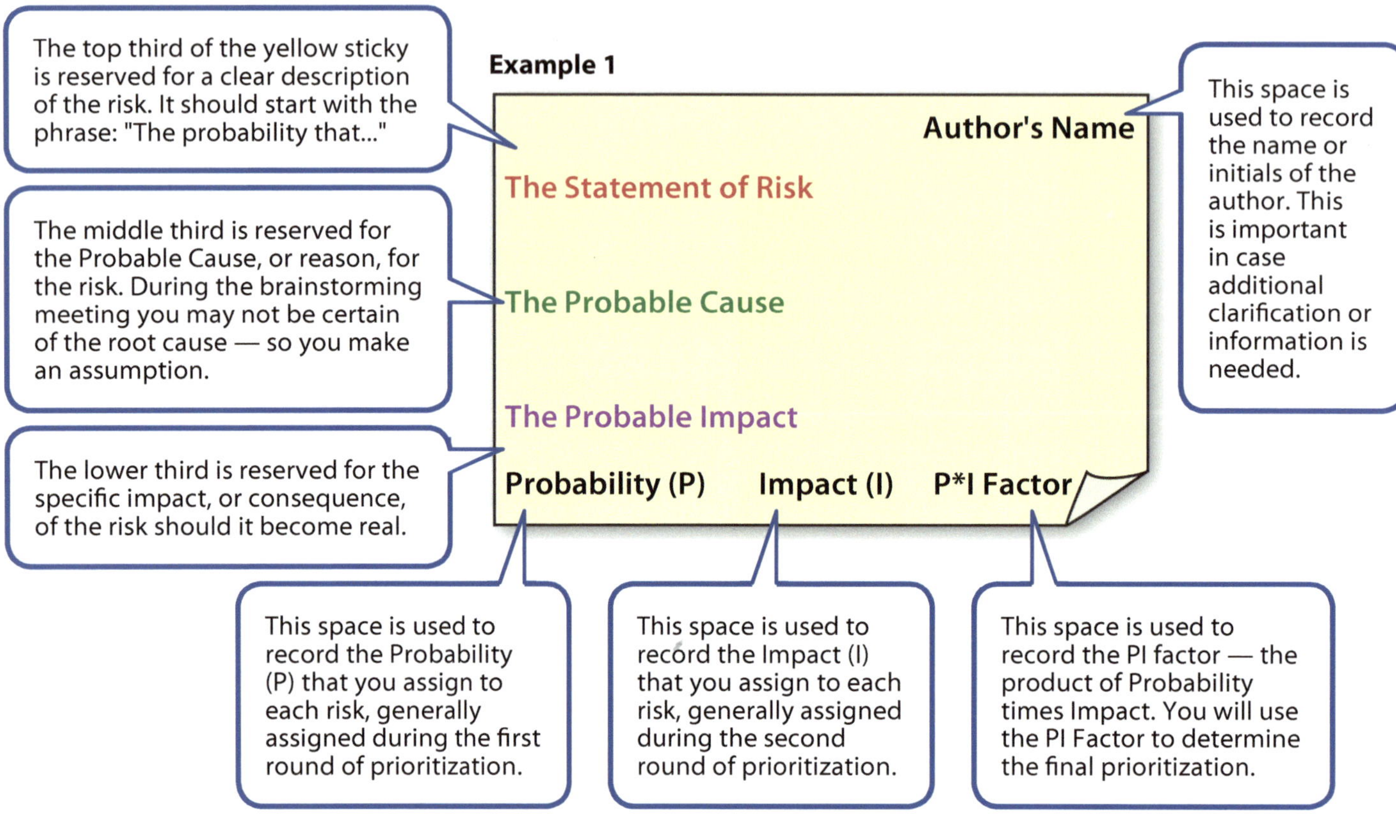

Example 2: The Incorrect Protocol

This is poor way to describe a risk. Note that the information is vague, it has no context; therefore, it can be interpreted in many different ways. You don't know who wrote it; consequently, it will be difficult to get a clarification. When someone writes a risk with too little information, help him rewrite it using the protocol.

Example 2

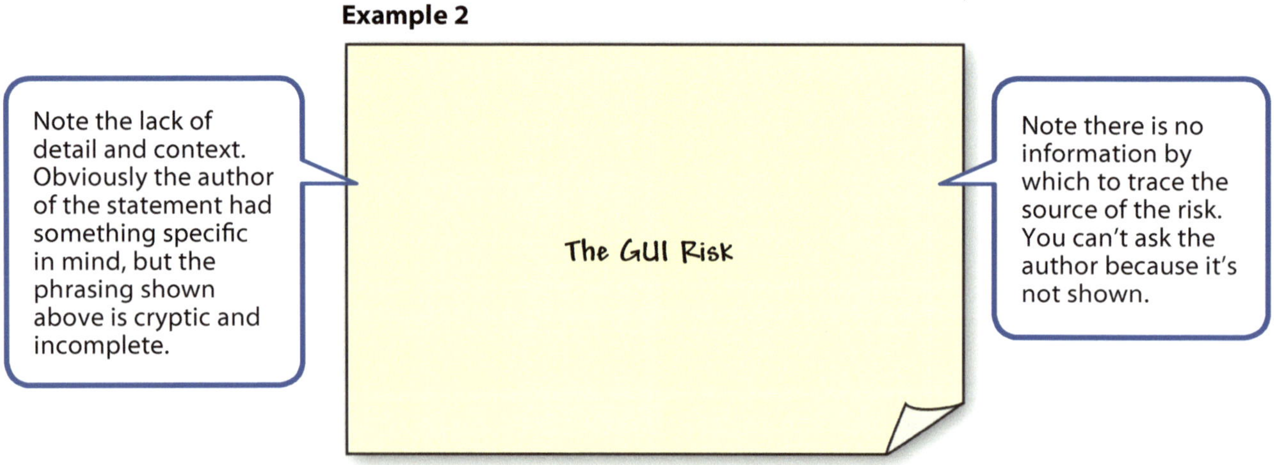

Example 3: The Correct Protocol

These examples illustrate the correct way to document a risk during your brainstorming session. The presence of the author's name allows for the opportunity to clarify. Also note that these examples have already been prioritized and include the PI Factor, which will enable the final prioritization.

Example 3-A

José C.

The probability that the new user interface does not meet customer requirements due to lack of usability research consequently making the new product difficult to use

P = 6 I = 8 PI = 48

Risk Statement

Probable Cause

Impact of Risk

Example 3-B

Mary M.

The probability that the new power supply will not be able to sustain the environmental requirements due to the low environmental ratings results in overheating causing catastrophic failure and potential for fire

P = 6 I = 8 PI = 48

Risk Statement

Probable Cause

Impact of Risk

Example 3-C

Jan J.

The probability that the main water pump will develop cracks due to the high temperature of the water, causing the cooling system to shut down, which would stop production

P = 6 I = 8 PI = 48

Risk Statement

Probable Cause

Impact of Risk

Identifying Risks

Example 4: Impact or Consequence?

In risk management, we always talk about the impact that a particular risk will have on your project. Impact is the negative effect that a risk has on the successful completion of your project. Consequence is another word for impact. Use either one. In FMEA, the preferred term is "impact."

Notice that "due to poor financing" is the possible root cause.

Example 4

> The probability that our main supplier in Singapore goes out of business due to poor financing.
>
> Causes a disruption in our JIT process, thus shutting down our manufacturing plant for three weeks.
>
> First session — Bill A.

Example 5: One Risk with Two or More Possible Impacts

In some instances, one risk can have two or more possible impacts. This is yet another reason for clarity in describing risks. It becomes essential; for example, consider the ramifications of two team members talking about the same risk, but having two different impacts in mind. Everyone must describe the risk using the protocol, to ensure absolute clarity.

Example 5-A

> The risk that our software is vulnerable to security attacks.
>
> Causes us to fail the acceptance test with our main customer, and thus the loss of a very large order.
>
> Session in Germany — Bill A.

Example 5-B

> The risk that our software is vulnerable to security attacks.
>
> Causes a major change in architecture, which results in a six-month delay to our rollout plan, and the loss of $1,000,000 in sales.
>
> Session in France — Bill A.

Example 6: The Wording of Risks and Potential Failures

The way a risk is stated is very important for the sake of clarity and also for processing the information. Below are additional examples that might help you articulate your risks.

Example 6-A

Notice that "due to the brittleness" is the possible root cause.

> Bending the power cord during manufacturing causes it to become frayed due to the brittleness of the sheath.
>
> Results in a safety risk that might produce electric shock to the point of severe injury to the customer.
>
> Session "A"; Group "2" Bill A.

Notice that in every case the impact is worded from the customer's point of view.

Example 6-B

> It is very likely that the radiator will leak after 12 hours of use, which would cause the engine to overheat and trip the temperature control shut-down relay.
>
> Results in making the vehicle inoperable for the customer.
>
> Session on the engine block Bill A.

In this example, the root cause is not stated; that is, the reason that the radiator leaks is not identified.

This is acceptable if the root cause will be addressed later in the process.

Notice that the code to trace the information is the risk management session.

Example 6-C

> There is a probability that the operator opens the safety cover and touches the high-voltage supply.
>
> Results in the death of the operator.
>
> 020507, 10AM Bill A.

This is clearly a show-stopper risk. Safety is always of extreme importance.

Identifying Risks

Example 7: A Chain of Risks

In some instances, one risk points the way for another risk. This is called a "chain of risks." This is a very important topic. Studies have shown that teams tend to overlook the risks caused by another risk-taking place. We suggest that you probe for additional risks that might be triggered by an initial risk.

Example 7-A

It is very likely that the radiator will leak after 12 hours of use, which would cause the engine to overheat and trip the temperature control shut-down relay.

Results in making the vehicle inoperable for the customer.

Engineer, EE — Bill A.

Example 7-B

Overheating the engine causes failure of the water pump due to the high temperatures of the water.

Results in making the vehicle inoperable for the customer.

Engineer, ME — Mary J.

Example 7-C

Failure of the water pump causes the engine block to crack.

Results in a very expensive repair bill for the customer.

Technician — Bob D.

Prioritizing Risks: Creating a List of the Vital Few

Introduction

Once you have identified all possible risks, it's time to prioritize them based on solid criteria that you and your team create. Prioritization can be achieved through the use of Affinity Diagrams. There is a full description and process map for Affinity Diagrams in the Critical Tools section of this guidebook. An additional method to prioritize is through a Karnaugh Map, which also is explained in the Critical Tools section of this guidebook.

The strategy for prioritizing is to enable your team to focus on the vital few; that is, those risks that are the most likely to occur in your project and can also have the most impact. The examples on the following pages show the mapping of risks using a Karnaugh Map.

Simply stated, not all risks have the same impact. Some can be very minor and insignificant, while others can indeed inflict major impacts on your project. By the same token, not all risks have the same probability of happening. Some are so remote that it's not worth worrying about them, while others have a high probability of happening. Consequently, it stands to reason that you want to sort risks by the highest probability of happening and by the highest impact. This combination provides the list of the vital few.

Once you have prioritized your risks, you need to develop your Risk Management Plan.

Karnaugh Maps (2x2 Matrixes)

Facilitators are often faced with helping a team make a decision or set priorities from a selection of many options. Generally, when choosing, we have two criteria — for example, Probable and Important. While some things can be very important, they are not necessarily probable. Thus, the ability of a team to separate all tasks based on the criteria will result in a better deployment of limited resources.

A Karnaugh Map is a participative tool that enables a group to prioritize or select items from a group. A Karnaugh Map is a 2x2 matrix that highlights how the criteria relate to an item and enables better and faster decisions to be made.

A PI matrix specifically highlights the probability and importance of a risk.

See Also: **Karnaugh Maps**

See Also: **Risk Management Plan**

LEVEL 1
The Master Process:
Developing and Implementing a Risk Management Plan

Phase 1: Assess

STEP 1. IDENTIFY RISKS

STEP 2. PRIORITIZE RISKS — You Are Here

Phase 2: Plan and Implement

STEP 3. PLAN PREVENTIVE AND CORRECTIVE ACTION (MITIGATE)

STEP 4. IMPLEMENT PREVENTIVE ACTION

Phase 3: Monitor and Control

STEP 5. MONITOR, MONITOR MORE, AND CONTROL

STEP 6. TAKE REMEDIAL OR CORRECTIVE ACTION

NEW TOPIC

Prioritizing Risks

Examples: Karnaugh Map 2x2 Matrixes

Each of the four corners represent the maximum value for at least one of the axes. In this case, the maximum (10) refers to the probability of the risk happening.

When the scales are the same (0 to 10), then this quadrant contains the most important risks; i.e., those with very high probability of happening and the most damage if they were to happen.

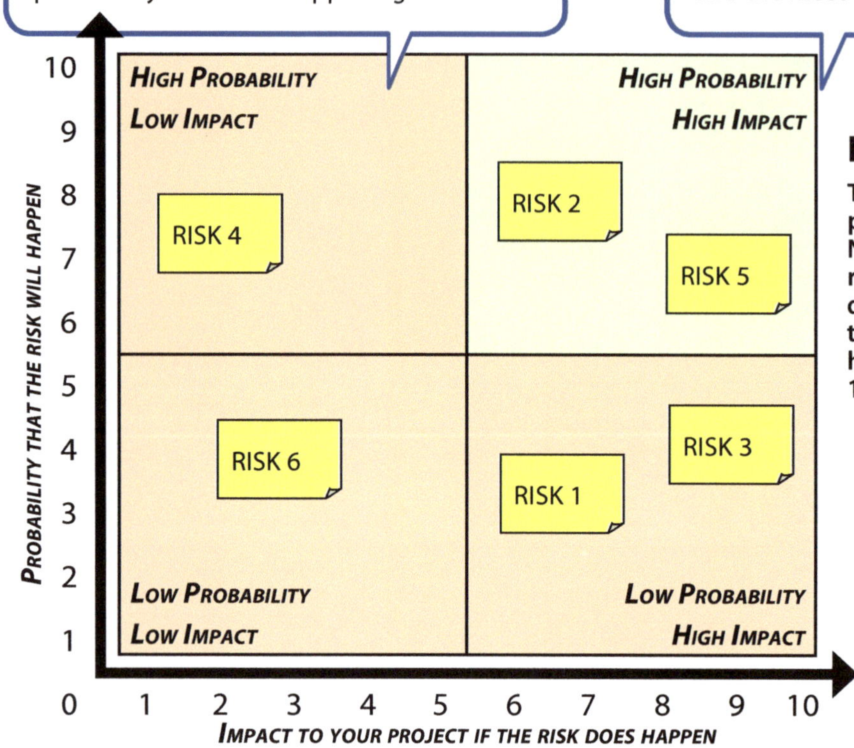

Example 1: PI Matrix

The y axis shows a scale for the probability of a risk happening. Notice the 0 to 10 scale, which is the recommended approach. The x axis depicts the impact; i.e., the damage to your product if the risk were to happen. Here the scale is also 0 to 10, 10 being the maximum impact.

This quadrant is the most important, because the risks are likely to happen, and there will be a serious impact if they do happen.

Example 2

This example demonstrates a team effort to prioritize risks associated with building a new house.

Notice the upper right quadrant where the most important risks are mapped — those risks have the highest possible impact and are most likely to happen.

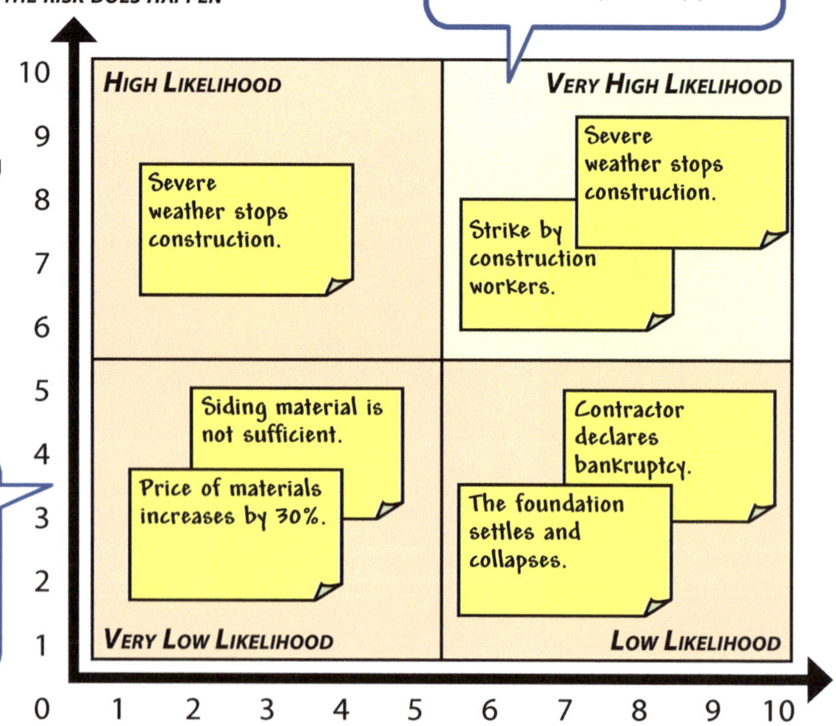

This is the "don't care" quadrant. Items in this area are not likely to happen, and even if they do, the impact will be miniscule.

Planning Preventive and Corrective Action

The Risk Management Plan

Definition

The mission of the Risk Management Plan (also called the Risk Response Plan) is to document and communicate risks, identify actions to reduce threats to the goals of the project and define clear responsibility for action.

The Risk Management Plan has at least four key sections:

1. **DESCRIBING THE RISKS IN ORDER OF PRIORITY** allows you to create the document of reference by which your team and managers can understand, identify and obtain the priority of risks for a specific project. It tells the reader which risks you have identified, which priority you have assigned and which risks you and your team have chosen not to track.
2. **PREVENTIVE ACTION** documents the specific steps you will take in anticipation to prevent the risk. In a best-case scenario, you can identify actions that will preclude the risk altogether.
3. **CORRECTIVE ACTION** documents the specific activities you will undertake should the risk occur.
4. **CLEAR RESPONSIBILITIES** documents who is responsible, when activities should happen and in what sequence.

What a Risk Management Plan Should Contain

There are certain elements that must be shown in order to define, track and manage risks. There are additional items contingent upon your circumstances. You must take into account the culture of your organization, the importance of the risk plan (every project is different) and the resources available to your team; for example, how many members and how easily can they meet. Use your judgment. Over time you will develop a Risk Management Plan that is unique to your organization and serves your objectives. Review the list on the next page to determine the elements you should include in your plan. There is flexibility; so, the following list is not rigid. The list simply reveals possible items in a Risk Management Plan.

In conclusion, your plan must contain enough information to enable your team to monitor and take action; in some companies the list of items in the plan is already defined. We suggest you review the list and ensure that it is still valid. Regardless, your plan must stand on its own, be clear and enable your team to implement it.

Essential

"Your plan must contain enough information to enable your team to monitor and take action."

Flexibility

The contents of your Risk Management Plan should reflect the culture of your organization, the objectives of your projects and your personal style. Be willing to experiment to find the best design for your company.

LEVEL 1
The Master Process:
Developing and Implementing a Risk Management Plan

Phase 1: Assess

STEP 1. IDENTIFY RISKS

STEP 2. PRIORITIZE RISKS

Phase 2: Plan and Implement

STEP 3. PLAN PREVENTIVE AND CORRECTIVE ACTION (MITIGATE) — You Are Here

STEP 4. IMPLEMENT PREVENTIVE ACTION

Phase 3: Monitor and Control

STEP 5. MONITOR, MONITOR MORE, AND CONTROL

STEP 6. TAKE REMEDIAL OR CORRECTIVE ACTION

NEW TOPIC

Planning Preventive and Corrective Action

Important Note

There are two different formats you can use to document a Risk Management Plan.

1. **THE TRADITIONAL, NARRATIVE VERSION:** Generally a written description of the key, prioritized risks and their corresponding mitigation and responsibilities. Many organizations no longer use the narrative way to document the plan; they prefer using a spreadsheet due to its many advantages.
2. **THE SPREADSHEET VERSION:** A spreadsheet serves the same purpose as the narrative version, but adds benefits, such as performing calculations, calculating dates dynamically, being stored in an electronic format that can be updated frequently and sent around the world.

It is up to you to decide which way you will document your plan. We suggest that you adopt the spreadsheet model.

Example: Possible Items in a Risk Management Plan

Item	Description
1. Work Package Number	In project management, it is the number of the specific task or work package.
2. Risk Event Number (Risk Number)	A sequential number assigned to every risk. Used simply for tracking purposes. In project management it can be the work package number, but you may choose to use a different sequence or format.
3. Risk Event (Risk)	The description in plain English of the risk at hand. For example, "Snowstorm prevents work." See Also: Protocol for Risk Definition
4. Risk Probability	The likelihood that a risk will occur. Expressed as a number from 1 to 10, 1 being the least likely and 10 being the most likely.
5. Risk Impact	Severity of its effect on the project objective. Also called "risk consequences" or "amount at stake." This item describes the damage if the risk becomes real. Expressed as a number from 1 to 10 where 10 equals the highest impact; i.e., more damage if the risk happens.
6. Risk Event Status	Measure of importance of a risk event. Also called criterion value or its ranking. Similar to "criticality," a cumulative ranking of a risk based on probability and impact.
7. Project Risk	The cumulative effect of the chances of uncertain occurrences adversely affecting project objectives.
8. Criticality	The overall importance (priority) of the risk when compared to all other risks in the plan. This can be expressed as a number; this number is very important, as it is used to prioritize all your risks in relationship to one another. Also called Risk Factor, you can multiply the risk probability times the risk impact, and the product of the two is the risk factor.
9. Impacted Risk	When one risk affects another risk, either positively or negatively. Also called a "risk chain."
10. Preventive	The actions you will take to prevent the risk from happening.

Planning Preventive and Corrective Action

Item	Description
11. Trigger Point	The date or other signal that you will use to deploy the preventive step. Yet another action or step you will take if your "Preventive" is not available. Some call this "Plan B."
12. Owner	The person who will monitor and take action on the risk at hand. The individual who is accountable for managing a specific risk or risks.
13. Risk Event	The description of what might happen to the detriment of the project.
14. Contingency Reserve	The amount of money or time normally included into the project cost or schedule baseline to reduce the risk that the project objectives exceed an acceptable level.
15. Management Reserve	The amount of money or time not included in the project cost or schedule baseline that is used by management to allow for unexpected and unpredictable situations.
16. Risk Financial Impact	A dollar estimate of the impact of a risk if it does happen. It can also signify the cost of preventive action. This is a very important number, as it assigns a meaningful figure (dollars) to your risk plan. Similar to Criticality but expressed in terms of financial impact.

This table contains a list of items, or topics, covered in a typical risk management plan. Your company may already have its own list, in which case you need to follow it. Not every item in the above list must appear in your plan. Rather, include every item that your team needs to monitor and take action on. You can collaborate with your team and decide the items, sequence and priority so it becomes a team list.

Risk Management Plan (RMP) as a Spreadsheet

Think of documenting your Risk Management Plan (RMP) in the form of a log using a spreadsheet. This format does away with excessive narrative and allows for more focused management of your plan.

Using a spreadsheet such as Excel is a convenient way to document your RMP. You can use the spreadsheet to capture, prioritize and monitor. Once you are ready, you can also document all the additional information you need to manage risks through the life of your development project.

Keep in mind that in a spreadsheet you may have several hundred risks captured, and when you are ready to document your plan in a spreadsheet, you have already identified, prioritized all the known risks using the PI factor (the product of the Probability times the Impact that you and your team designated to each of the identified risks). This happened in the brainstorming session you had with your team at the start of the process.

The PI Factor can be used in your spreadsheet to prioritize (for example, in Excel you can sort by the column that contains the PI Factor); this will rank each risk starting with those with the highest PI factor.

Once you have prioritized, you and your team need to decide how many of the risks you will manage. Use your judgment and remember that every development project is different; consequently, your choice will vary by project. For example in one you may chose to manage the top 25, while another may only require the top 10.

The information contained in a Risk Management Plan:

- The updates to the documented risks as more information becomes available.
- The dates where the review of the Risk Management Plan took place.
- New and emerging risks as information becomes available.
- Changes to the risk plan, for example responsibilities, course of action, etc.

Planning Preventive and Corrective Action

- Documenting when a risk actually took place.
- Deletion of risks that will no longer happen or have become "issues."
- Other information that helps you track, and monitor your Plan.

Be Proactive

The Risk Management Plan is essential for proactive project management; it is the process of controlling risk. It calls for a high level of discipline and mental concentration. It calls for much initiative on the part of the project manager. It calls for an instigator, or activist, who is resourceful and takes action toward eliminating or mitigating the risk.

General Rules about the Risk Management Plan

1. Ensure that regular progress meetings are established to identify and mitigate risk. Ideally, this is a separate and independent meeting, which allows sufficient time and concentration to the critical task of managing risk.
2. Keep your Risk Management Plan up to date; while it may be a burden, it will save considerable grief and aggravation. Include the Risk Management Plan as part of your regular project review meetings or status meetings; this way you will be forced to keep it up to date.
3. Don't wait until the next regularly scheduled meeting to address a risk. Rather, call an urgent meeting, or even an informal meeting, to address the challenge.
4. Define the action needed to address the challenge. It is not enough to meet and talk about the risk. It is critical that you insist on identifying the action needed to address the risk. Always use the following checklist: **DO WHAT? BY WHOM? BY WHEN?**

Issue Versus Risk

Some organizations use the terms "issue" and "risk" interchangeably, but this is incorrect. The Issue Log tracks issues in the project, including current and future issues. Future issues constitute risk. That being said, the Issue Log and the Risk Management Plan can be combined into a single document and a single meeting.

- **RISK:** Something that has a probability, though not 100 percent, of happening.
- **ISSUE:** Something that has a 100 percent certainty of happening.

Therefore, an issue is something you must deal with, as it is an absolute certainty that it will happen. Stop thinking about it, and take action!

Essential

"An issue is something you must deal with, as it is an absolute certainty that it will happen."

Planning Preventive and Corrective Action

Documenting Your Risk Management Plan

In this guidebook we provide two approaches to documenting your risk management. Your choice is based on your objectives. We provide examples and templates of both viable approaches.

Approach 1: Traditional Table

The traditional method has been to use a simple table to document your plan. The advantages of this approach are the flexibility to send the plan around the world, the ability to update the plan as needed and the ability to insert considerable information directly into the document or as a hyperlink to other documents.

Example: Documenting Your Risk Management

Risk Management Plan

Work Package or Task	Risk Event No. and Description	Probability	Impact	Plan			Actual			Owner
				Risk Event Status (P x I)	Criticality	Impact on Other Risks	Preventive	Trigger Point	Contingent	
Develop blueprints	Draftperson not available due to illness	2	7	14	Critical	Yes Engineering	Hire another draftperson	Draftperson calls in sick	Borrow draftperson from another department	Fred Smith
Stone work completed	Stone mason from Europe unable to arrive on time	5	6	30	Critical	No	Coordinate travel arrangements well in advance	The stone mason is not on site 6 days prior to start	Hire local stone mason to start positioning the material	May Jones

Approach 2: Electronic Spreadsheet

A spreadsheet such as Microsoft Excel is another fine way to document your plan. The advantages of this approach are the ability to perform calculations, the sort command for instant ranking, plus the benefits of an electronic document that can be sent around the world and frequently reviewed and updated.

You can use the spreadsheet to capture and then prioritize risks. You can also document your management plan by entering the information next to each of the prioritized items in your spreadsheet. After quickly calculating the PI Factor, you can use then use Excel's Data Sort feature and instantly prioritize the risks. In a spreadsheet you may have several hundred risks captured, yet easily manage all of the data.

Important: If you have the print version of this guidebook, e-mail the authors for a copy of the Microsoft Excel file: risk@rapidinnovation.com.

Implementing Preventive Action

NEW TOPIC

Plan for Success and Failure

At this point you have completed all your planning; your plan is documented and ready for action. Naturally, the first thing to do is implement all the preventive action contained in your plan.

As you recall, mitigation of risk consists of two parts. First take action to ensure that the risk does not happen, and second take action if the risk does happen. Preventive action is all about taking the steps needed to ensure that risks do not happen.

Preventive action should be documented in your plan, and you should not wait for implementation. Your best course is to ensure that the risks do not take place.

LEVEL 1
The Master Process:
Developing and Implementing a Risk Management Plan

Phase 1: Assess

STEP 1. IDENTIFY RISKS

STEP 2. PRIORITIZE RISKS

Phase 2: Plan and Implement

STEP 3. PLAN PREVENTIVE AND CORRECTIVE ACTION (MITIGATE)

STEP 4. IMPLEMENT PREVENTIVE ACTION (You Are Here)

Phase 3: Monitor and Control

STEP 5. MONITOR, MONITOR MORE, AND CONTROL

STEP 6. TAKE REMEDIAL OR CORRECTIVE ACTION

Proactive Risk Management

Once you have developed your Risk Management Plan, you move to Phase 3: Monitor and Control. In this phase, you closely monitor what is happening with your project and make adjustments to your plan based on the changes and progress.

This is where the concept of proactive risk management shines. Proactive risk management means that your plan is not stored in a filing cabinet or a hard drive, never to be seen again (which substantially reduces the benefits of the planning effort).

Instead, you schedule regular meetings to monitor and control risks; for example, once per week, once per month or at each project milestone. The meetings should be frequent enough to avoid missing a trigger point. The frequency of these meetings varies depending of the type of project you are managing — use your judgment. The important principle is to have continuous monitoring and control of risks.

Generally these are 60-minute meetings, scheduled solely for the purpose of risk management. We highly recommend that meetings be set to review and manage risks independently of other meetings. We have found that appending risk management to other meetings such as status meetings or project review meetings tends to shortchange the topic. However, you can schedule your risk management meetings right after or just prior to other meetings to take advantage of attendance.

Sometimes it's not practical to have a special meeting; some in your team may feel that there are too many meetings already. Then make your risk review part of a different meeting — for example, the regular project status meeting, or even a design review meeting. Make certain that you ensure sufficient time for risk management; you don't want to have the last five minutes of another meeting, this is not productive. Instead agree that you will have at least 30 minutes; better yet, take a break and let your team members rest. Then start again to talk only about risk management.

One last point: You need to invite key players to your monitor and control meetings — leaders and managers most intimate with the project.

Monitoring the Risks

Monitor Your Risk Plan Using the Critical Questions of Proactivity

Agenda for the Risk Management Meetings

The list shown below will provide a starting point so you can develop an agenda that is better suited to your projects. You can add, change or delete items to improve the productivity of your meeting.

We suggest that you project the current version of the Risk Management Plan so everyone can see the changes being made. If you do not have access to a projector, a paper copy of the plan will suffice.

1. What has changed since we last met?

- ❑ What important changes have taken place that impact our project?

2. How have risks changed?

- ❑ Are there any new risks that must be added to the Risk Management Plan?
  - What has caused the new risks?
  - What are the implications of the changes?
  - Who are the risk owners for the newly identified risks?
- ❑ Are there any old risks that are obsolete and must be deleted from the Risk Management Plan
  - What cause the deletion of risks?
  - What are the implications of the changes?
- ❑ What changes have happened to the current list of risks?
  - What caused the changes to existing risks?
  - Change of Risk Owners?
  - Update to the P and the I?
  - Any new data that add clarity to the risk?
- ❑ What other changes have take place?

3. What are the impacts to the overall Risk Management Plan?

- ❑ Are there any trends that will impact our Risk Management Plan?
- ❑ What will happen if the new trends continue? What should we do right away to mitigate the trends?

4. Validate the needed changes and then document them into the plan prior to the close of the meeting.

- ❑ What are the specific changes to our current Risk Management Plan?
  - Add new risks?
  - Delete risks?
  - Changes to current risks?
  - Changes in responsibilities?

Taking Remedial or Corrective Action

Taking the Next Steps

You and your team have been constantly monitoring the Risk Management Plan to identify any trigger points, define new risks and take preventive action to keep risks from happening.

At this point, you have documented what needs to be done in case a trigger occurs. Further, you know who is responsible and you have allocated sufficient resources to handle matters. In short, you are well prepared due to the risk planning you did earlier and your constant monitoring of risks.

When a risk does happen, you are ready to take remedial or corrective action. The purpose of such action is to ensure that your project delivers the anticipated results.

Address the Risks

As you monitor your Risk Management Plan, you will determine that a risk has reached the trigger point — the point documented in the plan at which mitigation will initiate. Take appropriate action, and ensure that those responsible for various activities know it. Your plan already has the specific action you need to take, but you may alter the plan depending on the situation.

Address Risks Not in the Plan

Regardless of the diligence in developing your plan, there is always the possibility that an unexpected risk happens. Because it is not contemplated in the plan, you will need to act fast. Focus on the risk, define a level-headed course of action and lead the implementation. This is where a Risk Management Team comes into play; your team will collaborate with you in addressing the risk and defining appropriate action.

LEVEL 1
The Master Process:
Developing and Implementing a Risk Management Plan

Phase 1: Assess

Step 1. Identify Risks

Step 2. Prioritize Risks

Phase 2: Plan and Implement

Step 3. Plan Preventive and Corrective Action (Mitigate)

Step 4. Implement Preventive Action

Phase 3: Monitor and Control

Step 5. Monitor, Monitor More, and Control

You Are Here

Step 6. Take Remedial or Corrective Action

PRODUCT DEVELOPMENT

FMEA: Failure Mode and Effect Analysis

Using FMEA in Risk Management 77

Using FMEA in Risk Management

NEW TOPIC

FMEA: A Framework of Risk Management

The Benefits of FMEA

The business case for deploying a framework of risk management is to achieve any of four specific goals:

1. Reduce time-to-market.
2. Improve reliability.
3. Reduce waste (improve efficiency).
4. Improve customer satisfaction.

The need to improve productivity, as defined by the four goals above, is an enduring effort of any product development company, particularly those that face global competition. The case can be made that engineering productivity will be enhanced by a model of risk management.

Definition

Failure Mode and Effect Analysis (FMEA)

A logical, structured analysis of a system, subsystem, device or process used to identify possible failure modes, their causes and the effects of these failures, which will result in solutions that increase the overall reliability and safety of current and future products (Schubert, 1992).

Simply put, FMEA is a structured and disciplined way to go from symptom to remedy.

Several Applications of FMEA

1. Analyzing potential reliability problems early in the development cycle, when it is easier to take actions to overcome these issues — thereby enhancing reliability through design.
2. Identifying potential failure modes, determining their effect on the operation of the product, and planning actions to mitigate the failures. A crucial step is anticipating what might go wrong with a product. While anticipating every failure mode is not possible, the development team should formulate as extensive a list of potential failure modes as possible.
3. It can also be applied for corrective action: If the device fails, FMEA is used to identify the reasons for failure, initiate preventive measures to avoid the failure reoccurring and mitigation steps to remedy it if it does reoccur. Generally, this type of FMEA is done as part of debriefs or retrospects; service and repair organizations also take advantage of it.

Important

FMEA has application anywhere there is a desire to manage risk or important decisions need to be made.

Early and consistent use of FMEA in the design process allows the engineer to design out failure and design in reliable, safe and customer-driven products. FMEA also captures historical information for reliability analysis for use in future designs.

FROM SYMPTOM →	TO DIAGNOSIS →	TO REMEDY
Effect on the Customer (Impact or Severity)	Failure Mode Root Cause	Corrective Action

The above diagram maps the conceptual approach to FMEA, where the starting point is a symptom; for example, "The door does not close, causing hot liquid to spill on the operator." This is followed by the diagnosis, which in some cases takes the form of Root Cause Analysis. The conclusion is the remedy, or corrective action, which is the action needed to eliminate the root cause to prevent the risk from happening.

Using FMEA in Risk Management

The Business Case for FMEA

The business case for a model of risk management (FMEA) can be made by focusing on four goals:

1. Reduce waste (improve efficiency).
2. Reduce time-to-market.
3. Improve reliability.
4. Improve customer satisfaction.

See Also Business Case for Risk Management

FMEA is Intended for Documentation:

1. A failure or potential failure.
2. The modes of the failure.
3. The effects of the failure.

Through Rigorous Analysis:

1. What happens.
2. How it happens.
3. Why it happens.
4. What to do about it so it does not happen again.

When to Apply FMEA

1. At the start of the product design process, perhaps even prior to the design, to evaluate new technology or devices.
2. When changes are made to the operating conditions of the product or when performance changes are implemented.
3. When the design of the product is changed.
4. When the design process itself is changed.
5. When a new technology is incorporated into an existing or new product.
6. When new regulations or industry standards are implemented.
7. When service tickets go out of tolerance.
8. When the product will be used in a different environment than that for which it was designed.
9. When new suppliers are hired.
10. When a new tool or manufacturing process is introduced.

Using FMEA in Risk Management

Types Of FMEA

1. Concept FMEA
   - Intended to minimize failure effects on the system or subsystem at the concept level; i.e., pre-design.
   - Assumes a failure, and then identifies how that failure could occur at the system level.
   - The functional approach that generally involves a top-down analysis in which a specific failure model for the entire system is traced back to the initiating subsystem's failure modes.
2. Design FMEA
   - Minimizes failure effect on the design. Intended to minimize failure in products before they are released to manufacturing.
   - Focuses on possible design deficiencies.
3. Process FMEA
   - Minimizes process failure's effect on the total process.
   - Used as a means to ensure that potential failure modes identified in the design process are not allowed during manufacturing, and to prevent new failure modes as a result of the manufacturing processes.
4. Service FMEA
   - Minimizes service failures.

FMEA is a Hierarchical and Bottom-Up Model

1. Product Level

2. System Level

3. Subsystem Level

4. Component Level

5. Nano Level

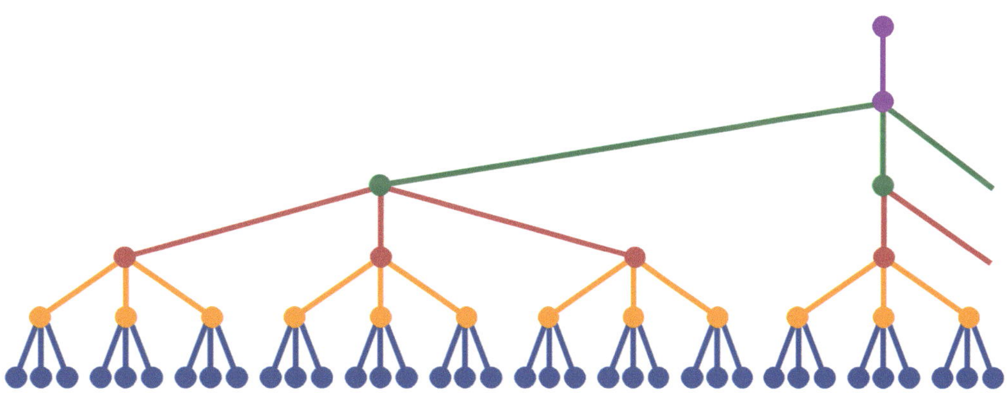

The above diagram is a depiction of the various levels of FMEA. You and your team must decide which level is needed to meet your objectives. Remember that FMEA is a bottom-up methodology. The strategy is to do the analysis starting at the lowest logical level and then work your way up to satisfy your objectives.

Using FMEA in Risk Management

The Attributes of FMEA

1. **PROACTIVE:** It encourages corrective action before a failure occurs.
2. **DISCIPLINED:** It provides a rigorous framework from end to end.
3. **SYSTEMATIC:** It implements methods for:
   - Analyzing potential failure modes (recognizing, describing and evaluating).
   - Ranking the risks associated with failure modes.
   - Identifying root causes.
   - Prioritizing them for remedial action.
   - Defining corrective action to prevent and mitigate failures.
   - Acting on the highest-ranked items (planning, measuring and implementing).
   - Reevaluating those items and returning to the prioritization step in a continuous loop until a marginal return sets in.
   - Documenting and archiving for reference and future applications.

Flexible versus Restrictive FMEA

Generally, FMEA is voluntary; i.e., although your company may require it or your engineering group decides it in necessary, it has not been imposed by an external organization or entity. When it is voluntary, you have as much flexibility as you need. You can change the process, add steps and create new forms as long as you preserve the initial intent of FMEA.

On the other hand, there are situations where a customer requires you to do FMEA. In that case, your flexibility is limited; you have to follow the guidelines established by the customer. Be certain to ask for documentation such as definitions, expected documentation, reporting, etc. In addition, ensure that the method of FMEA you intend to use is the same as that of your customer's. In theory, industry-standardized FMEA should be the same; however, some organizations add their own requirements.

There is another case where your flexibility will be limited, and that is liability. Consult with your legal department to define if the FMEAs you do are subject to possible legal liability. If the answer is "YES," then you have a new set of requirements to meet (too long for this guidebook to cover). We suggest you consult a reliable book on FMEA such as *Failure Mode and Effect Analysis, second edition* by D. H. Stamatis, published by ASQ Quality Press.

Types of FMEA

The core of all FMEA is the same: The intent is to identify failure modes in a disciplined and traceable way. FMEA has become a popular methodology, used beyond product design. Several types include:

- System: Focus on global system functions.
- Design: Focus on components and subsystems.
- Process: Focus on manufacturing and assembly processes.
- Service: Focus on service functions.
- Software: Focus on software functions.

FMEA is, after all, a form of risk management; consequently, it has application anyplace where there is a desire to manage risk or where important decisions need to be made.

Using FMEA in Risk Management

FMEA and Legal Liability

There are situations where the probability of liability may be high; that is, your company may be liable for damages if something goes wrong with your product. For example, in the automotive, aerospace, naval and other industries liability is a serious concern. You must consult your legal advisors and executives to understand the potential for liability. This will significantly impact the FMEA you perform.

Laws vary around the world, so this author believes that it is prudent that you investigate thoroughly the implication of liability for your company and industry. Investigate the local laws and tax codes of the places where you do business. You must clearly understand the implications of liability prior to performing an FMEA.

Suggestions for further study on the implications of liability on FMEA:

- Consult your legal advisors. Many corporations have lawyers and legal experts who can help you. The executives of your company can also be a source of valuable information, but keep in mind that they may not be fully aware of all the implications of the law in a particular country
- For further reading, we suggest *Failure Model Effect Analysis* by D. H Stamatis. Mr. Stamtis dedicates a full chapter to the issue of liability.

See Also Recommended Books

FMEA and Military Standards in the USA (MILSPECS)

Military organizations around the world have requirements for risk management, particularly FMEA. You may find yourself having to comply with a military FMEA. Obtain the specific document from that military organization and follow the specification faithfully, particularly the type of documentation that you are asked to provide.

For the military in the USA there are several general specifications: **MIL-8TD-785B**, **4MIL-STD-1629A** and **ARP-926A**.

You can get information about FMEA for the military at the Society of Reliability Engineers website: www.sre.org. Specific FMEA standards for military can be found at Weibull, a private company: www.weibull.com/knowledge/milhdbk.htm#400. In the USA, two other websites that can be helpful: www.dsp.dla.mil and https://assist.daps.dla.mil/quicksearch. Finally, professional associations generally have reference materials on military specifications for FMEA; for example, IEEE (Institute of Electrical and Electronic Engineering) www.ieee.org and the Society of Automotive Engineers www.sae.org

CAUTION! Military standards are numerous and can change. You should verify that the standard you have been given is appropriate. Do the proper research to ensure that you are using the right documentation.

FMEA and Project Management

In project management you perform risk management on your project. While not exactly a FMEA, the two do share many common elements. The basis in project management is to define, manage and mitigate risk. In project management there is a tool called the PI matrix, which is remarkably similar to the FMEA matrix. Not surprising, because both can be thought of as risk management tools. If you have project management training, you already have a solid foundation and should find FMEA easy to understand and use.

Using FMEA in Risk Management

Risk Management Strategy Matrix for a Development Program

Template

The following matrixes compare types of FMEA with design phases — each cell is an opportunity for FMEA. Only cells that represent particular areas of risk need to be filled in. You can use this framework as a template to map your FMEA strategy for a new design. Use your judgment, guided by the business objectives of your project, to determine the number, type and timing of your FMEA.

Example 1: Types of FMEA (Y) Versus Program and Project Levels (X)

Types of FMEA	Program Level	Project Level
Financial (Business Case)		
Marketing (The Market Place)	The Product Marketing Manager shall develop a Risk Management Plan to address and manage risk. Emphasis on our efforts to enter the new market.	
Engineering Design (Software and Hardware)		
Manufacturing (Process)		The Manufacturing Manager will develop a Process FMEA to address and manage risks associated with the new manufacturing process. Emphasis on those parts that will be outsourced.
Supply Chain (Suppliers)		
Service and Repair (Field service, repair and maintenance)	The Service Manager shall develop a Risk Management Plan. Emphasis on the warranty period, and service for countries in Asia.	

CONTINUED

Using FMEA in Risk Management

Example 2: Types of FMEA (Y) Versus the Design Phase (X)

Types of FMEA	Level of the Risk Management Plan During the Product Life Cycle				
	Product Definition	Product Design	Product Rollout	Sustaining Phase	Pre-Phaseout
System level		Engineering will develop and track a risk management plan (RMP) to address risks in system integration. Emphasis will be placed on outsourced subsystems.			
Subsystem level					
Component level					
Nano level					
Process FMEA			Manufacturing in collaboration with engineering will develop and track a Process FMEA. Emphasis on the new manufacturing and test process.		
Software level					
Service level		Service will address reliability issues particularly in DOA and develop steps to prevent or mitigated risks.			

Using FMEA in Risk Management

FMEA Glossary

Term	Definition
Cause	• The reason why a particular element of a design or process results in a failure model. It is the reason for the failure, as defined through Root Cause Analysis.
Component Function	• The function as it was intended to perform. This is a short description of the intended design of the component and its performance. You can create a link to another document with additional information; for example, schematics, theory of operations, bill of materials, etc.
Component Description	• It is a short description of the component, subsystem or system where the failure may occur. So your team can better understand, you can cross-reference to schematics, other documents, theory of operation, etc.
Corrective Action/ Recommended Corrective Action	• The approach you define to fix the problem. Remember that there is corrective action — the long-term fix for the problem — and remedial action — the temporary, band-aid fix. • There are several means to fix a failure; for example, poka yoke to foolproof, redesign the part, improve reliability, etc.
Criticality	• A form of FMEA that emphasizes the impact and probability (severity and occurrence) to characterize the impact of the failure on the customer. For example, "severe injury to operator" and "90 percent impact (severity)" indicates a very serious consequence to the customer.
Current Controls	• The name of the person who is most likely to be responsible for detecting or triggering the root cause of the failure; for example, a mechanical engineer. • It may not be possible to identify a specific person, in which case assign it to someone on a temporary basis. There may be several persons who can detect the failure or its root cause, in which case identify a leader and record the other names.
Detection	• The identification of a failure cause or its consequent failure mode prior to the delivery to the customer. Thus, it is your ability to detect the failure before it reaches the customer. • However, sometimes it is the ability of the customer to detect the failure. You and your team must agree on one approach. The traditional approach is your ability to detect it. • An assessment of the likelihood that current controls will detect the failure mode or its effect. The ease of detecting the failure is indicated by a numeric scale, generally between 1 and 5.
Detection Ranking	• A ranking, based on a predefined scale, of the probability of your operation's ability to detect failure before the customer. Some failures are difficult to detect and others are obvious — this is the ranking. There are many scales, but in FMEA 1 to 5 is the most common.
Effect	• The impact on the customer; that is, the means by which the failure shows up to your customer. The corresponding impact. The way the customer describes the problem.
Failure Mode	• The way in which the device fails to meet its intended function. The way in which a product or process failure manifests itself to the customer.

Using FMEA in Risk Management

Term	Definition
Function	• The function of the part, subsystem or system to be analyzed; i.e., the task that the system must perform. (It is implied that the system refers to the system subsystem, part, process or even service for which the FMEA will be performed.)
Occurrence ("P" for Probability)	• The number that indicates the frequency of the potential failure. The assessment of the probability that a particular cause will happen and result in a failure mode. • Occurrence in risk management is "P" for probability.
Poka Yoke	• Japanese for "error-proofing." A quality strategy emphasizing the prevention of defects at their source by making changes that prevent them from happening.
Potential Failure Mode	• The way in which the device may fail to meet its intended function.
Potential Cause	• The possible cause of the failure.
RPN (Risk Priority Number)	• A number computed by multiplying the occurrence, severity and detection values (OxSxD= RPN). The greater the number, the greater the concern. • It is used to arrive at a way to prioritize failure modes — one way to reduce all the potential failure modes to the vital few. • It is critical for you and your team to agree on the criteria that you will assign to the RPN; that is, an RPN is a number, but your team must give it a meaning that represents an agreed-to priority, not just the numerical or absolute value of the RPN.
Severity ("I" for Impact)	• A number, generally between 1 and 10, that indicates the severity of the potential failure. It is an assessment of how serious the impact of the effect is on the customer. • The severity scale must be agreed to by the team prior to and during the FMEA process to ensure clarity. There are accepted scales for certain industries, but it is critical for the team to define and agree on the meaning of the scale to define severity. • Severity in risk management is "I" for impact.

Using FMEA in Risk Management

Detailed Process: FMEA

Important

Because not every FMEA is the same — components are different, as are market conditions, urgency, etc. — it's worth spending a bit of time defining some basic assumptions; for example, why are you embarking on an FMEA? It is a healthy and worthwhile exercise to lay a solid foundation.

We recommend as an initial phase, Staging your FMEA. Staging will allow you to prepare for the task, answer critical questions, align managers and executives, etc. Ultimately, staging will help you make the best use of the resources, commitment and dollars you will invest into the FMEA.

See Also: Map Your FMEA Strategy

See Also: Types of FMEA

See Also: Team Charter

1

Staging Your FMEA

Your objective is to schedule, plan, and facilitate a brainstorming session to get as much input as possible from as many stakeholders as possible. Because this meeting is the first tangible manifestation of your efforts, you must prepare to ensure a successful event. Here's a check list:

- Identify the people who will be invited to participate in the first brainstorming meeting, ensuring that there is diversity of experiences, functions and perspectives. Naturally, those involved in the project are a "must invite." Invite your Risk Management team if you have one.
- Get a commitment from each guest that they will participate for the entire duration. This first meeting is bound to be long, from four to eight hours. (Eight hour meetings can be split into two four hour sessions.) Inform your guests so they can agree to attend for the duration.
- Select the type of FMEA you will perform.
- Define and document the reason why you are about to perform a FMEA.
- Ensure the support of the appropriate supervisors, managers and executives, as the FMEA will require resources.
- Clearly identify the part, subsystem or system for the FMEA.
- Prepare the venue, supplies, samples, documentation and other supporting materials that will aid in performing a successful FMEA.

Other documentation that might help:

- Block Diagram
- Design drawings.
- System schematics and other schematics.
- Functional diagrams.
- Previous data.
- Manufacturer's components.
- Service records.
- Calibration records.
- Customer complaints.
- Previous ECNs.
- ISO-9000 CARs.
- Detailed bill of materials.
- Theory of operation.
- List of components.
- Actual samples of the device.
- Actual samples of devices that have already failed.

These lists are, of course, do not constitute a comprehensive listing. Use your judgment to decide what will be helpful. Talk to the members of your team, and ask for input on additional documentation.

Important

Staging should include the completion of the Team Charter — a written record of your decisions. It engages your management chain. More important, it adds considerable clarity to the serious endeavor of risk management.

Continued

Using FMEA in Risk Management

Cont'd

2

Confirm Your Objectives

You must confirm that your objectives are still valid — strong enough to justify a FMEA — as it takes time and resources; in some cases, it adds time to your already tight schedule.

Key questions to ask:

- Why should we do this FMEA?
- What specific results do we expect? For example, improved reliability, reduction of time-to-market, improvement of safety, etc.
- What is the approach (strategy) we should select? For example, "We will perform a system-level FMEA," "We will perform a subsystem-level FMEA with special focus on safety," and "We will perform an FMEA with special focus on serviceability."

Also part of your strategy is timing. Generally, a FMEA should be performed at the design stage, but even in this area there are phases. You could do a FMEA during the development of the technology, at design completion. The rule of thumb is it's never too early to start your Risk Management Plan!

3

Identify the FMEA Team

This is a critical step.

Ensure that the team is clearly cross-functional and multi-disciplined.

Ensure that every team member is willing to contribute and understands the importance of your FMEA. Avoid recalcitrant or unwilling participants.

Validate with supervisors and managers that employees will have the time and support to help you with the FMEA.

Important

It's critical to have a diverse group of people on your FMEA team. A homogenous group tends to have a limited view of potential failure modes. Invite people from other parts of the organization who can add value. For example:

- Procurement and purchasing.
- Quality management.
- Metrology.
- Service, repair and calibration.
- Manufacturing outgoing QC.
- R&D.
- Your customer.
- Field engineering.
- Overseas personnel and suppliers.
- Documentation customer training.
- CAD and CAM people.
- Outside consultants, experts, university professors or industry experts.
- Internal experts from other parts of the company.
- Your outsourcing partners.

Time Frame

Due to the complexity or scope of the FMEA, you may want to break up the meeting into several sessions. For example, do the brainstorming in one three-hour session, and the rest in a subsequent session a day or two later. This avoids fatigue and gives the team the opportunity to introspect.

Continued

Using FMEA in Risk Management

CONT'D

4

SCHEDULE THE FMEA BRAINSTORMING MEETING

This meeting must be carefully planned. Generally, it should take more than three hours, and in some cases much more. Some FMEAs may take a full day.

Ensure that the room is large and appropriate for teamwork. Have ample quantities of office supplies such as yellow stickies, felt-tip markers, flip charts, etc.

Important

Be prepared at the start of the meeting to answer your team's questions; for example:

- What is the timeline for this FMEA?
- What is going to happen to the information the team generates?
- Do we have a budget for research and analysis?
- Do we have the full support of management?
- Can we access other resources? Even if they cost money?
- Why are we doing this? I am too busy doing real work. Is this really important?

WHERE YOU'RE AT

Now that you have done the necessary preparation, it is time to begin the FMEA Brainstorming Meeting.

5

START THE FMEA BRAINSTORMING MEETING

Start by clearly stating the objective of your FMEA. Ensure that everyone in your meeting understands the objective by allowing sufficient time for discussion and questions.

Your team members will be more productive if they understand the "whats" and "whys" of your FMEA. They need to know what prompted you to call a FMEA meeting. They also need to know why they are on the team. They need to know what you expect from them, in no uncertain terms, during and after the meeting. Ensure clarity!

This section of your meeting is to set the context; i.e., the big picture. Do not rush. It may take 30 to 60 minutes to add the needed clarity. But it's worth it!

6

DEFINE THE FUNCTION OF THE PRODUCT OR PROCESS

FOR A PRODUCT: A block diagram or theory of operation will be very helpful, as will an actual sample. The intent is to describe the function of the device or product such that everyone in the room has a reasonable understanding of what, how and why it does what it does. In some cases, the understanding needs to be in-depth. Naturally, this will take longer, and the people in the room must have the skills and abilities to understand.

Remember, not everyone in the meeting is as familiar with the product as you are; further, some in the meeting may not be as technically proficient as others. Take the time to ensure understanding to the level where they can make a contribution later in the meeting.

FOR A PROCESS: Use a process map or a flow diagram. The purpose of this step is to ensure that everyone on your team understand the product or process under consideration. Do not shortchange this step — some may not be familiar enough to contribute, and this step will provide needed understanding. This is a good opportunity to also describe the problem (failure mode). Stay with it until everyone shows a reasonable understanding of the matter at hand.

NOTE: Define the conditions of the function; for example, "under laboratory conditions," or "at the customer's site," or "in the manufacturing line" or "under extreme environmental conditions." And, for each, you need to provide a criteria; for example, temperature, pressure or voltage. Publish your plan as appropriate.

CONTINUED

Using FMEA in Risk Management

Cont'd

7

Explain the Yellow Sticky Protocol

It is critical that you capture the information from the meeting participants; consequently, you must enforce a protocol for capture and documentation of the information.

Everyone must have a pad of Post-It Notes® (trademark of 3M) and a Sharpie® (Sanford brand, fine point, black markers).

Ask everyone to write down — and continue to write down — potential failures. Everyone must use the protocol.

Everyone must read each yellow sticky (failure) aloud so everybody can hear.

After reading the yellow sticky, post it on the wall. You may have several hundred by the time you are done

See Also Protocol for Risk Definition

8

Identify Potential Failure Modes

The best way to get things started is to ask: What could go wrong? What did go wrong?

Here is where the investment you made in the team pays off.

This is a brainstorming step; thus, do not limit the conversation. Allow for all ideas to flow, even ones that seem off-the-wall or flaky. The ideas will be refined later. What are **ALL** the ways in which the service or product could fail?

Important

Defining the function of the product and identifying the potential failure modes can be done as discrete steps, or you can encourage your team to articulate the failure and the severity as one sentence. The protocol for yellow stickies assumes that both are stated together. We recommend that people think about the failure and the severity together.

However, it is not wrong to do these steps separately. You could even mix it up: some members of your team separate these steps, and some combine them.

Where You're At

- You have explained the reasons for your FMEA session.
- You have explained the function of the device, system, subsystem for which you are performing the FMEA.
- You've allowed time for the participants to ask questions and fully understand the function of the device.
- You've asked the participants to write down every possible failure mode.
- The participants have written the failure modes using the Yellow Sticky Protocol.
- There are several hundred yellow stickies

9

Scrub, Clean Up and Thin Out

You may have several hundred yellow stickies pasted on the wall. Many of them could be duplicates, redundant or simply not valid. You and your team should spend time ensuring that all the yellow stickies are relevant.

As you review each note for validity, do save the yellow stickies you remove.

This would also be a good time to rewrite those yellow stickies that are not clear enough. Ask the author to better explain the information and write a new yellow sticky.

Important

This may be a good time let your team rest (coffee or tea break). They have probably spent several hours identifying risks. In some cases, you may want to call a second meeting to finish the FMEA.

Continued

Using FMEA in Risk Management

CONT'D

See Also **Ranking Scales**

10

RANK (PRIORITIZE) YOUR FMEA

For each failure mode (Risk) assign a number from 1 to 10.

Write the number on the yellow sticky using the protocol.

Document your scale for future reference.

1. **RANK THE SEVERITY (S) OF EACH FAILURE MODE**
   The more severe the impact of the failure, the higher the number.
2. **RANK THE OCCURRENCE (O) OF EACH FAILURE MODE**
   The higher the probability (occurrence) of the failure, the higher the number.
3. **RANK THE DETECTABILITY (D)OF EACH FAILURE MODE**
   The more difficult it is for you and your team to detect the failure, the higher the detectability number.

YOUR CHOICE

A — PERFORM ROOT CAUSE ANALYSIS (RCA) NOW

B — PERFORM ROOT CAUSE ANALYSIS (RCA) LATER

CRITICAL DECISION

At this point of your FMEA you need to make a critical decision. There are two paths available to you, and they depend on your objectives and the approach (strategy) you have selected.

CHOICE A: Most textbooks will ask you to identify the root cause of every potential failure after you assign the severity scale. There is much value in this approach, as it forces you to address all potential failure and identify all root causes. This is important when you need to "ruggedize" your design or product, when there is a high probability of legal liability, or when you have a strong need to improve reliability. The downside is that it is tedious. Also, you have to do root causes for items that have a very low probability of happening.

CHOICE B: The other path is to postpone identifying the root cause. After assigning the severity scale, assign occurrence, continue sequentially to compute the RPN and, finally, identify root causes. The advantage of this path is that you only do Root Cause Analysis on failures that have a high RPM number; i.e., the most critical failures. The disadvantage is that you may miss an opportunity to improve reliability, as some failures will not be addressed.

Remember, it's your choice. You and your team should have a good conversation about it and decide.

11

COLLECT ALL INFORMATION

After you have assigned the Criticality, Occurrence and Detectability to every yellow sticky, you need to collect all the information and get it ready to take with you.

Collect all the yellow stickies, flipcharts, etc. You will need all this information for the continuation of your FMEA efforts.

12

CONCLUDE THE BRAINSTORMING MEETING

You can dismiss the team.

This phase of your FMEA is finished.

For comments, questions or additional copies please e-mail the author: risk@rapidinnovation.com.

Using FMEA in Risk Management

CONT'D

13

CREATE THE SPREADSHEET

In the past, this phase of your FMEA was done manually. Spreadsheet software, such as Excel, reduces the stress and accelerates the process.

There are examples of the spreadsheet in this guidebook, including Excel demos in the interactive PDF version. If you did not obtain the interactive PDF edition, please send a message to risk@rapidinnovation.com.

See Also
Documenting Your Risk Management Plan
Approach#2: Electronic Spreadsheet

14

ENTER THE DATA

Enter the information from each of the yellow stickies, including the three rankings: Severity (S), Occurrence (O) and Detectability (D).

This may be a tedious task, particularly if you have several hundred yellow stickies. It may be helpful if you recruit several members of your team to share the job.

15

CALCULATE THE RPN (RISK PRIORITY NUMBER)

Used in FMEA, the RPN is the product of the Probability (S), the Occurrence (O) and the Detectability (D); SxOxD = RPN.

The result is a number that allows you to rank the risks based on an overall priority once you have determined the RPN for all risks. This is useful when there are many risks in your FMEA. It is a simple, linear multiplication of three values. The higher the number, the more priority a particular risk has.

Jan J.

The probability that the main water pump will develop cracks due to the high temperature of the water, causing the cooling system to shut down, which would stop production

S = 7 O = 4 D = 2

By now, you will have a set of yellow stickies, perhaps as many as 100, or even more.

Each yellow sticky will have three numbers at the bottom for each of the three variables needed to calculate your RPN.

16

RANK FAILURES BASED ON THE RPN

After you have entered all the data from all the yellow stickies, you can calculate the RPN.

Sort your spreadsheet data by RPN to quickly see the rank of all the failure modes.

17

SET THE CRITERIA

You and your team must make a decision: How low in the ranking will you do Root Cause Analysis? The top five? The top 20? Use your judgment. You need not do all the failures, as the lowest ones are often irrelevant. You might want to use the Pair-wise approach to help narrow your choice.

Do not write off the lower-ranking failures. You may choose to do Root Cause Analysis on some of the lower-ranked failures because you have a strategic reason to do so; for example, it will improve reliability, design, manufacturability, service or repair.

CONTINUED

Using FMEA in Risk Management

CONT'D

WHERE YOU'RE AT

DECIDE IF YOU CAN IMPLEMENT CORRECTIVE ACTION

Now that you have ranked every yellow sticky using the RPN number and have decided how many failures to address, you need to analyze each of the prioritized failures.

- Is it clear enough for full understanding?
- Do we have enough information to proceed with Corrective Action?

Important: For every failure there is a root cause — generally not the detection (symptom), but the actual technical cause of the failure. There are a number of tools available to help you and your team perform a RCA. Some of those tools are included in this document:

- Fishbone Diagram.
- Affinity Diagram
- The 5 Whys.

There are many other tools available to solve difficult, technical Root Cause Analysis such as Triz. Your company may already have some processes in place, or you can research the available choices.

YOUR CHOICE

A — INSUFFICIENT INFORMATION

Cannot define corrective action.

B — NO NEED FOR ADDITIONAL DATA

The corrective action is obvious and the current information is sufficient.

18 ASSIGN RESPONSIBILITIES FOR THE ADDITIONAL RESEARCH

19 PERFORM THE ADDITIONAL RESEARCH

This is an action item. The proper detection methods are to experiment, collect data and test for validating the root cause of the failure.

In some cases, this will take days, perhaps weeks — the data collection and diagnosis can be time-consuming.

You may need an action plan to ensure completion.

20 ASSIGN RESPONSIBILITY FOR CORRECTIVE ACTION

You have concluded the FMEA (for now), and now you are ready for action. At this point you need to create an action plan that includes all the failures that you have selected for corrective action. Determine who is responsible for the action and when the action should be completed.

Additional meetings will be required for your team members to report progress and to guarantee closure.

Additional FMEAs may be required as the design progresses or as new potential failures are identified.

Important: Corrective action, as per ISO-9000, is the permanent solution implemented as a result of the Root Cause Analysis.

Remedial action is the temporary solution you implement, commonly called the "band-aid solution." It does not remove the root cause of the problem.

The purpose of FMEA is to implement corrective action.

CONTINUED

Using FMEA in Risk Management

CONT'D

21

ESCALATION

You now have a long list of risks and, not surprisingly, some of the risks impact other functions or programs. Perhaps some of the risks are more strategic; others have a serious financial impact.

You can create a separate list and escalate them to the proper levels of management in your organization. Perhaps some need to go to another function, such as marketing, finance or sales.

Refrain from disposing of these homeless risks, because they can impact your organization.

STOP

LEARN AND IMPROVE (KAIZEN)

This step is part of your continuous improvement efforts. Your FMEA process will improve significantly if you constantly add lessons learned.

After your project is completed, identify how well your FMEA worked.

Learn from the experience — document and use lessons learned.

Build a Best Known Practices (BKP) document for the FMEA so that you and others will have the information available for the next project.

Using FMEA in Risk Management

Template

The FMEA Management Plan

The initial Brainstorming meeting is but the start of the FMEA process. After you've identified and prioritized the possible failure modes, you need to document and also create an Action Plan in order to monitor progress and ensure results. Start the plan immediately after the Brainstorming meeting to capture the information — view the section on Risk Management Plan (RMP) as a Spreadsheet in this guidebook for additional information.

See Also
Risk Management Plan (RMP) Spreadsheet

Example 1: Basic FMEA Management Plan

FMEA Management Plan										
Tracking Number	Statement of Failure (Risk)	Possible Root Cause	Statement of Impact	Occurrence (Probability)	Severity (Impact)	Detectability	RPN	Preventive Action	Mitigation	Owner
N	The seal in the connection between the main pump and the intake hose breaks	Due to deficient environmental specifications of the seal	Causes the pump to run dry and burn out bring the process to a complete halt	7	9	8	504	Specify a new seal with stronger environmental specs.	Increase the testing of the seal. Test every connection	Fred A.

Above is a typical example of a spreadsheet you develop to act as your FMEA plan. Naturally, other categories can be added, plus links to other documents. You and your team need to define the best content for your FMEA plan.

Information in the FMEA Management Plan

The table below describes the most common item you might want to include in your FMEA Management Plan. This list is not all-inclusive. Work with your team, and define the information you need to monitor and manage your FMEA. Be careful not to add more information than you need.

Each category in your list could be entered in a spreadsheet column. You could even add to your spreadsheet links to documents such as schematics, data, etc.

Category	Description
Tracking No.	A code or number for reference purposes that helps you track and manage your data. This number of used only for tracking purposes, it serves no other function. The tracking number should be sequential and formatted in a way that you and your team believe helpful.
Component Description	A short description of the component, subsystem or system where the failure may occur — be sure to clearly describe the item being evaluated.

CONTINUED

Using FMEA in Risk Management

Category	Description
Statement of Failure (Failure Mode or Risk Description)	A narrative description of the risk. It must be clear and sufficient such that others in your team can understand it. This description is essential to the success of your FMEA. This guidebook provides a protocol for describing risks.
Potential Root Cause	The description of the potential root cause of the Statement of Failure. During the brainstorming meeting you team can speculate about the root cause of any failure mode. It is not until you have done Root Cause Analysis that you can categorically state the cause.
Statement of Impact	The consequences or damage that the risk would cause should it take place.
Occurrence Ranking (Probability)	The likelihood that the specific risk will occur. By definition a risk has a level of uncertainty; the probability is your attempt to assign a value to it. In FMEA the scale is always from 1 to 10.
Severity Ranking (Impact)	The consequence. That is, the damage or negative results if the risk actually happens. In FMEA the scale is generally from 1 to 10. The damage is always described from the customer's perspective; the impact that the failure would have on the user or customer.
Detectability Ranking (Detection)	Your ability to detect a failure before the product reaches your customer. How obvious or how difficult it is for you and your team to detect the manifestations of a failure. In FMEA the scale is generally from 1 to 5.
RPN	RPN (Risk Priority Number) is the product (multiplication) of the Probability, the Impact and the Detectability (SxOxD = RPN) (used in FMEA). The result is a number that allows you to rank the risks based on an overall priority.
Mitigation	The specific steps to prevent a risk, or the required corrective action if it does happen.
Trigger	The flag or signal you will use to launch your mitigation steps.
Risk Owner	The person responsible and accountable for monitoring a specific risk. The entire team relies on this person to keep an eye on the risk.
Date of Last Review	The date when a particular risk was reviewed, updated and discussed. The risk owner is responsible for ensuring the review, updating the team and discussing the risk.
Status at Time of Last Review	A brief narrative of the status of a particular risk. It could be the termination of a risk as proper disposition has occurred. The status narrative should be clear such that the entire team can understand the current situation of the risk.
Date of Actual Impact	Allows you to track the actual date when the failure manifested itself. It may be helpful for reliability analysis.
Actual Recovery Taken	The actual mitigation taken. It may help you compare the planned action to the actual mitigation.
Current Control	The name of the person who is most likely to detect the root cause of the failure; e.g., a mechanical engineer. It may not be possible to identify a specific person; in such case, assign it to someone on a temporary basis.

CONTINUED

Using FMEA in Risk Management

Example 2: Expanded FMEA Management Plan

This example was derived from actual FMEAs. Obviously, you will need to adapt it to meet your business objectives and the culture of your team. When defining your categories, we suggest that you capture only the information you need to meet your objectives. Also, there may be specific guidelines as to the type, sequence and formatting of the information imposed on you by your company, customers or regulatory agencies; investigate to ascertain if any rules are in place for your FMEA.

FMEA Management Plan

Tracking Number	Component Description	Failure Mode (Risk Description)	Occurrence Ranking (Probability)	Severity Ranking (Impact)	Detectability Ranking	RPN	Mitigation	Trigger	Risk Owner	Date of Last Review	Status at Last Review
1	Cooling lines	The risk that, due to a customer change, all water lines in the cooling system require double containment (which we do not have now), which causes the customer to reject the equipment.	5	9	3	405	Complete redesign of the cooling system.	Customer brings it up during program review meeting or writes an RCN (Requirement Change Notice).	Fred S., Cooling System Design	October 15	No change.
2	Motherboard	The risk that EMI is generated, which degrades the deflection electronics, which causes us to not meet the published specifications.	7	9	9	567	Add more safety reviews; include more testing in the test plan.	EMI is detected during testing.	May L., reliability manager	October 7	A plan for measuring EMI was delivered by Roger F. on October 6.

Note: To preserve the linearity of the scales, 10 is very difficult to detect; thus, in this example a Detectability Ranking of 3 means that it is easy to detect and a Detectability Ranking of 9 means that it is very difficult to detect.

Examples of Ranking

Ranking the Probability of Failure

In the following pages we provide actual examples of ranking for Occurrence, Detection and Severity. The tables below outline different ways to assign probability. These are not universal standards of ranking, but they may assist you in developing the ranking you will use.

Remember, there is a difference between **ACCURACY** and **RESOLUTION**. There is a risk of having very high resolution, and yet adding little if any accuracy. Use your judgment when designing your scale.

The most important factor is that you and your team have as much clarity as possible on the scale and its definition. That being said, don't fall into the trap of analysis paralysis, spending too much time on minutia or discussing a resolution that may not be necessary for your particular FMEA.

The following examples are grouped into three sections: **OCCURRENCE**, **DETECTION** and **SEVERITY**.

Example 1: Occurrence, Probability or Likelihood

Rank	Scale Definition
1	Remote probability of occurrence. Unreasonable to expect failure to occur. 1 in 1,500,000.
2-3	Low failure rate/probability of occurrence. Unreasonable to expect failure to occur. 1 in 1,500,000.
4-6	Moderate failure rate. Related to similar designs that have occasional failures of the moderate type. 1 in 2000, 1 in 400, 1 in 300.
7-9	High failure rate. Relates to failures in similar designs that have failed in the past. 1 in 20, 1 in 8, 1 in 3.
10	Very high failure rate. Almost certain that failure will take place in a significant way. 1 in 2 or greater.

Using FMEA in Risk Management

Example 2: Occurrence or Likelihood Ranking Criteria (Probability)

Probability of Failure	Failure Probability	Rank
VERY HIGH: Failure is almost inevitable.	>1 in 2	**10**
	1 in 3	**9**
	1 in 8	**8**
	1 in 20	**7**
HIGH: Repeated failures.	1 in 80	**6**
	1 in 400	**5**
	1 in 2,000	**4**
LOW: Relatively few failures.	1 in 15,000	**3**
	1 in 150,000	**2**
REMOTE: Failure is unlikely.	<1 in 1,500,000	**1**

Example 3: Occurrence or Likelihood Ranking Criteria (Probability)

Likely Failure Rates over Life of Device		Probability of Failure
1	< 0.1 per thousand devices.	**REMOTE:** Failure is unlikely.
2	0.1 per thousand devices.	**LOW:** Relatively few failures.
3	0.5 per thousand devices.	
4	1 per thousand devices.	**MODERATE:** Occasional failures.
5	2 per thousand devices.	
6	5 per thousand devices.	
7	10 per thousand devices.	**HIGH:** Frequent failures.
8	20 per thousand devices.	
9	50 per thousand devices.	**VERY HIGH:** Persistent failures.
10	> 100 per thousand devices.	

Example 4: Occurrence or Likelihood Ranking Criteria (Probability)

Rank	Description
1	**AN UNLIKELY PROBABILITY OF OCCURRENCE** during the item's operational time interval. Unlikely is defined as a *single failure mode (FM) probability <0.001* of the overall probability of failure during the item's operating time interval.
2-3	**A REMOTE PROBABILITY OF OCCURRENCE** during the item's operating time interval (i.e., once every two months). Remote is defined as a *single FM probability >0.001 but <0.10* of the overall probability of failure during the operating time interval.
4-6	**AN OCCASIONAL PROBABILITY OF OCCURRENCE** during the item's operating time interval (i.e., once a month). Occasional is defined as a *single FM probability >0.01 but <0.10* of the overall probability of failure during the operating time interval.
7-9	**A MODERATE PROBABILITY OF OCCURRENCE** during the item's operating time interval (i.e., once every two weeks). Moderate probability is defined as a *single FM probability >0.10 but <0.20* of the overall probability of failure during the operating time interval.
10	**A HIGH PROBABILITY OF OCCURRENCE** during the item's operating time interval (i.e., once every two months). High probability is defined as a *single FM probability >0.20* of the overall probability of failure during the operating time interval.

Example 5: Occurrence or Likelihood Ranking Criteria Probability

Level	Likelihood	Probability of Occurrence
1	Not Likely	~10%
2	Low Likelihood	~30%
3	Likely	~50%
4	Highly Likely	~70%
5	Near Certainty	~90%

Using FMEA in Risk Management

Example 6: Occurrence Ranking According to U. S. Military Standard MIL-STD-1629A

Level	Name	Description
A	Frequent	A high probability of occurrence during the item operating time interval. High probability may be defined as a single failure node probability greater than 0.20 of the overall probability of failure during the item operating time.
B	Reasonably Probable	A moderate probability of occurrence during the item operating time interval. Moderate probability may be defined as a single failure mode probability of occurrence greater than 0.10 but less than 0.20 of the overall probability of failure during the item operating time.
C	Occasional	An occasional probability of occurrence during item operating time interval. Occasional probability may be defined as a single failure mode probability of occurrence greater than 0.01 but less than 0.10 of the overall probability of failure during the item operating time.
D	Remote	An unlikely probability of occurrence during item operating time interval. Remote probability may be defined as a single failure mode probability of occurrence greater than 0.001 but less than 0.01 of the overall probability of failure during the item operating time.
E	Extremely Unlikely	A failure whose probability of occurrence is essentially zero during item operating time interval. Extremely unlikely may be defined as a single failure mode probability of occurrence which is less than 0.001 of the overall probability of failure during the item operating time.

The above table was reproduced from the U.S. Military Standard: MIL-STD-1629A.

Using FMEA in Risk Management

Example 7: Detection Ranking Criteria

Level	Probability	Rank	Severity of Failure as Seen by the Customer
10	Almost impossible to detect	Absolute certainty of non-detection.	Cannot detect or is not checked.
9	Very remote	Current controls will probably not detect.	Control is only achieved with indirect or random checks.
8	Remote	Controls have poor chance of detection.	Control is only achieved with visual inspection.
7	Very low	Controls have poor chance of detection.	Control is only achieved with double visual inspection.
6	Low	Controls may detect.	Control is achieved with charting methods such as SPC Control Chart.
5	Moderate	Controls may detect.	Control is based on GO/NO gauging performed on 100% of the parts.
4	Moderately high	Controls have a good chance to detect.	Error detection in subsequent operations.
3	High	Controls have a good chance to detect.	Error detection in station.
2	Very high	Controls almost certain to detect.	Error detection in station causes automatic stop.
1	Certain	Controls certain to detect.	Item has been error-proofed by design.

Example 8: Detection Ranking Criteria

Rank	Description
1-2	**VERY HIGH** probability that the defect will be detected. Verification and/or controls will almost certainly detect the existence of a deficiency or defect.
3-4	**HIGH** probability that the defect will be detected. Verification and/or controls have a good chance of detecting the existence of a deficiency or defect.
5-7	**MODERATE** probability that the defect will be detected. Verification and/or controls are likely to detect the existence of a deficiency or defect.
8-9	**LOW** probability that the defect will be detected. Verification and/or controls not likely to detect the existence of a deficiency or defect.
10	**VERY LOW** (or zero) probability that the defect will be detected. Verification and/or controls will not or cannot detect the existence of a deficiency or defect.

Using FMEA in Risk Management

Example 9: Detection Ranking Criteria

Detection	Likelihood of Detection by Design Control	Rank
Almost impossible to detect	Design control cannot detect potential cause/mechanism and subsequent failure mode.	10
Very remote	Very remote chance the design control will detect potential cause/ mechanism and subsequent failure mode.	9
Remote	Remote chance the design control will detect potential cause/mechanism and subsequent failure mode.	8
Very low	Very low chance the design control will detect potential cause/mechanism and subsequent failure mode.	7
Low	Low chance the design control will detect potential cause/mechanism and subsequent failure mode.	6
Moderate	Moderate chance the design control will detect potential cause/mechanism and subsequent failure mode.	5
Moderately high	Moderately High chance the design control will detect potential cause/ mechanism and subsequent failure mode.	4
High	High chance the design control will detect potential cause/mechanism and subsequent failure mode.	3
Very high	Very high chance the design control will detect potential cause/mechanism and subsequent failure mode.	2
Certain	Design control will detect potential cause/ mechanism and subsequent failure mode.	1

Example 10: Severity or Impact Ranking Criteria (Impact)

Effect	Severity of Effect	Rank
Almost impossible to detect	Very high severity ranking when a potential failure mode affects safe system operation without warning.	10
Very remote	Very high severity ranking when a potential failure mode affects safe system operation with warning.	9
Remote	System inoperable with destructive failure without compromising safety.	8
Very low	System inoperable with equipment damage.	7
Low	System inoperable with minor damage.	6
Moderate	System inoperable without damage.	5
Moderately high	System operable with significant degradation of performance.	4
High	System operable with some degradation of performance.	3
Very high	System operable with minimal interference.	2
Certain	No effect.	1

Example 11: Severity or Impact Ranking Criteria (Impact)

Level	Rank	Severity of Failure as Seen by the Customer
1	None	No noticeable defects.
2	Very minor	Abnormal noises, noticeable only by the most discerning operators.
3	Minor	Abnormal noises, noticeable by 50% of the customers. Fully operational, but causes stress on the operator.
4	Very low	Abnormal noises, noticeable by the customer. Fully operational, but causes stress on the operator.
5	Low	Operable but difficult to operate, all functions available.
6	Moderate	Operable but difficult to operate, some functions not available.
7	High	Operable, but at reduced level of performance.
8	Very high	Device inoperable.
9	Hazardous with warning	Unsafe to operate.
10	Hazardous without warning	Operator injury very likely.

Using FMEA in Risk Management

Example 12: Severity or Impact Ranking Criteria (Impact)

Rank	Description
1-2	Failure is of such a minor nature that the customer (internal or external) will probably not detect failure.
3-4	Failure will result in slight customer annoyance and/or slight deterioration of part or system performance.
5-7	Failure will result in customer dissatisfaction and annoyance and/or deterioration of part or system performance.
8-9	Failure will result in high degree of customer dissatisfaction and cause non-functionality of system.
10	Failure will result in major customer dissatisfaction and cause non-system operation or non-compliance with government regulations.

Example 13: Severity or Impact Ranking Criteria (Impact)

Rank	Scale Definition
10	Failure could injure the customer.
9	Failure would create non-compliance with regulation.
8	Failure renders the unit inoperable; total shut-down.
7	Failure causes a high degree of customer dissatisfaction.
6	Failure results in a subsystem malfunction, which creates product malfunction.
5	Failure creates customer complaints.
4	Failure can be addressed and fixed by the customer.
3	Failure can create minor customer dissatisfaction, but performance will not be compromised.
2	Failure is not visible to the customer, but will have some impact on the customer.
1	Failure will not be noticeable by the customer, and will not impact the customer.

Example 13: Severity or Impact Ranking Criteria (Impact)

Rank	Severity Level	Description
10	Catastrophic I	A failure results in the major injury or death of personnel.
7-9	Critical II	A failure results in minor injury to personnel, exposure to harmful chemicals or radiation, a fire or a release of chemicals into the environment.
4-6	Major III	A failure results in a low-level exposure of harmful chemicals or radiation to personnel, or activates facility alarm system.
1-3	Minor IV	A failure results in minor system damage, but does not cause injury to personnel or allow release of chemicals into environment.

Example 15: Severity or Impact Ranking Criteria According to MIL-STD-1629A

Category	Name	Description
I	Catastrophic	A failure which may cause death or weapon system loss.
II	Critical	A failure which may cause severe injury, major property damage or major system damage, which will result in mission loss.
III	Marginal	A failure which may cause minor injury, minor property damage or minor system damage, which will result in delay or loss of availability, or mission degradation.
IV	Minor	A failure not serious enough to cause injury, property damage or system damage, but which will result in unscheduled maintenance or repair.

The above table was reproduced from the U.S. Military Standard: MIL-STD-1629A

Critical Tools

Deployment of a New Tool. 107
Deploying a tool takes extra work. You are not just enabling the adoption of a new tool or method, but actually changing old habits. Don't be daunted by setbacks; you're doing the right thing for your team and company.

Problem-Solving Process 114
Don't underestimate the need to apply a formal process to solving a problem — simple or tough!

Karnaugh Maps: 2x2 Matrixes 119
Help your team make a decision or a selection to set priorities from an array of many options; for example, determine the risk factors of a new deployment or select key features on a new product.

The 5 Whys 125
This is a simple methodology to obtain valuable information from customers in order to get to the root message.

Yellow Sticky Protocol. 127
This is a way to track and code information to enable future processing. It's very useful when you have a large number of stickies or many people contributing.

Affinity Diagrams 130
We are not linear thinkers. We tend to jump around when brainstorming. Using Affinity Diagrams is a simple method for organizing the brainstorming session — the non-linear flow of ideas — into useful, documented facts. And it is a fine way to ensure consensus.

Prioritizing Affinity Diagrams 134
Prioritizing is all about making decisions and reaching consensus. The secrets are selecting the right criteria to use in making those decisions and clear communication.

The Fishbone Diagram 137
It's tempting to try to solve the symptom of the problem without trying to analyze the real cause. This is a fine tool for providing the discipline to go from "symptom" to actual "root cause" of the problem.

Cascading a New Tool to the Organization

Introduction

There is a difference between attending a seminar or workshop and causing your organization to embrace any or all of the tools you learned. Said another way, just because you and a number of your peers attend a seminar, that's no guarantee that what you learned will become common practice in your organization.

Deploying a tool takes additional work. It is a process, and requires planning, tenacity and leadership. Consider that you are not just enabling the adoption of a new tool or method, but actually changing old habits. In short, you will cause disruption and a break from the comfort zone currently in existence in your organization.

The Business Case

The return on investment (ROI) from any seminar or workshop only occurs after the class and when visible and measurable improvement in productivity is achieved. Some have the erroneous notion that attending a seminar is the objective; instead, attending is a means to and end. The real objective is to take what was learned and apply it to improve productivity.

Expect Resistance to Change

However, just because you and a number of your peers attend a seminar or workshop, that's no guarantee that what you learned will become common practice in your organization. Deploying a tool takes additional work. It's a process that requires planning, tenacity and leadership. Consider that you are not just enabling the adoption of a new tool or method, but actually changing old habits. In short, you will cause disruption and a break from the comfort zone currently in existence in your organization.

Expect resistance to change in overt and passive ways when you suggest the adoption of a new tool or method. You will hear things like "It won't work here," "We tried that, and it didn't work," "Our situation is different," and many similar statements. There are numerous resources about driving change in an organization to help you.

The Hesitaters

The initial push-back may actually be the expression of other feelings:

- **SEEKING ADDITIONAL INFORMATION:** Some in your team resist as a way to obtain additional information, as they feel fearful to deploy something they do not understand. It's the best way they know for doing fact-finding.
- **SEEKING REASSURANCE:** They feel unsafe; for example, they want to know that they will not get fired if the tool does not work the first time out.
- **TESTING:** They are simply trying to determine if this is for real and if you have a strong conviction that the tool will work.
- **GENUINE CONCERN:** Some may actually feel that the tool will not work or is not applicable.

The way to deal with all the above feelings is to first be aware that they are common, and then validate the actual source of the concern by providing additional information, reassurance, etc.

> **Important!** Much that has been written about driving change in an organization. When searching for books, seminars, articles and other resources use the terms "Managing Change" and "Resistance to Change."
>
> Also, your human resources department may be able to help or offer additional resources.

The Reactionaries

Some individuals are simply opposed to change, reactionaries who want to preserve the status quo. Occasionally you will run into someone who will refuse to change, refuse your new tool and steadfastly remain opposed. It is rare that you are able to change these individuals, and we don't have the time here to delve deeply into this topic. We suggest that you do not become discouraged; rather, build a constituency of those who do want to change, and use it to overcome the resistance.

Food for Thought

The reluctance to change is a part of the human condition; you are not a bad person if you resist change. Your strategy to overcome resistance must take into account that it is a normal reaction. Further, the process to overcome resistance is all about people. Do not rely on templates, software programs, process maps, etc. Rather, seek to understand the situation, and then tap your leadership skills to drive the change.

Here are a few things to keep in mind:

- First, set about understanding the current situation of those who need to adopt the new tool or method. Talk to them as a group and, more importantly, individually. Make an effort to listen with the intent of understanding.
- Be sure to consider WIFM (pronounced "wiffem"), which stands for "What's in it for me?" We react positively when we understand our personal benefits. You must make the case for embracing the change such that every person on the team sees his or her benefits as part of the deal.
- Stop selling and start communicating! Our excitement for the new tool gets the best of us, and we start pushing the idea to the point where it becomes annoying. Instead, remain open to suggestions. Some will try to help you succeed by providing ideas for improvement — for example, a better way to use a template or a better way to deploy the tool. Be open to ideas, and engage those who provide them; do not reject them either directly or passively.
- Speak their language. Adapt your message to the audience. Speak in a way that is familiar to them. This will improve communication and avoid the use of new words.
- Start small. The intent of the first use of your new tool or method is to train and to develop the organization; thus, do not select a major project. Rather, find a simple project where the use of your new tool will shine and the risk of failure is very small.
- Finally, your tenacity and leadership will carry the day. Don't be disappointed with setbacks. In the end you're doing the right thing for your team and your company.

The Critical Questions

All of us need a certain amount of information before we commit to a tool, method or practice. It is not fair to expect people to jump at the opportunity if they don't have sufficient context, especially at the personal level.

Generally, we all want the following questions answered:

- Why are we doing this?

  (What is the relevance to my team, our company or our organization?)
- What would happen to us if we don't do it?

  (What would be the impact on or consequences to the company, organization or team?)
- Why are you asking me to do this?

  (Tell me why I should care. What's in for me?)
- What support can I expect from you?

  (Can I rely on you? What is your commitment to me?)
- What specifically do you expect me to do?

  (Tell me specifically how this impacts my job. What specific changes must I make?)

The above information must be communicated in the language and context of the listener. Avoid platitudes and irrelevant information as they tend to confuse and add uncertainty.

Deployment of a New Tool

Detailed Process: Deploying a New Tool

1

Choose the Tool

It's best to start small. Deploy a single tool to reduce the challenge and risk of failure.

Define clear criteria; for example:

- Ease of learning.
- Amount of leverage the tool will provide.

You can select any criteria, but you need to be consistent in your evaluation and prioritization.

Start recruiting supporters and allies who react positively to the new tool.

Important! Use the Planning Document in this section to ensure that you are well prepared for the deployment.

2

Identify Your Core Team

You will need a small group of your peers to help you deploy the tool.

Team members should be willing and able to spend the needed amount of time helping you with the task. Sometimes it may take up to four hours per week.

Make sure the team will be in place for at least three calendar months. It might take longer.

Important! Remember to secure the visible support of your organization's executives. Without it, you risk failure.

Meet with your executives, and show them the benefits and improvements that your new tool will bring to the organization.

If possible, secure a sponsor or a mentor — one of the executives who is willing to assist you in the deployment.

Seek support from others who are willing to help deploy.

3

Secure Support from Your Executives

It is prudent to identify one of the executives of your organization to sponsor your efforts.

Sponsors, also called Mentors, will provide moral support, help remove obstacles and add a measure of stature to your efforts.

The best sponsor is one who, like you, has a passion for the tool you are leading.

Important! As you select the pilot for your deployment, also ensure that you have training, templates and other needed support to help the team succeed.

Continued

Deployment of a New Tool

Cont'd

4

Identify the Pilot

This is the team, organization or group of employees that will apply the tool to a real project.

The results of this pilot will provide valuable information when deploying the tool to the entire organization.

Remember to obtain approval from the appropriate managers and supervisors.

Important

Equip the Pilot

It is critical that you provide coaching, encouragement and support to the pilot team. There is the risk of push-back and disappointment if the tool does not work as expected

It would be best if you were part of the pilot team. That way, you will be there throughout the process.

5

Conduct the Pilot

Ensure that you provide a high level of support to the organization such that the deployment of the new tool does not end in failure.

Provide all the training, tools, encouragement and other forms of support throughout the deployment of your pilot.

Important

What if the Pilot Fails?

Even the best of efforts may results in failure.

The first thing to do is to debrief and identify the root cause of the failure — without this step you cannot recover.

Avoid blaming, guilt or punishment as these things add no value to the recovery process.

Once you have the root cause and lessons learned, modify the process and try a new pilot.

6

Debrief the Pilot

The topic of the debrief is to identify what worked and what did not so you can make appropriate improvements for the full deployment.

This debrief is done with your Core Team.

7

Proceed with Full Deployment

Deployment of a New Tool

Planning Checklist

Deploying a tool is indeed a project. Consequently, ensure that you have a deployment plan, a schedule, budget, etc. Working with your Core Team, or with the peers who have agreed to help, you can develop a tactical plan that will ensure a successful deployment.

The list below is intended to help you formulate your own plan. You may need to add your own items to the list. Also note that the list below is in no particular sequence. Finally, the list is not rigid. Use your judgment and do what makes sense within your organization.

Planning Checklist

- ❑ Schedule for the deployment defined, documented and approved.
- ❑ Training materials for employees developed and documented.
- ❑ Training materials produced — hard and soft copies — available for use.
- ❑ Training materials placed in the intranet or web page.
- ❑ Trainers of the new tool identified and informed of their duties.
- ❑ Training of employees scheduled and employees notified.
- ❑ Training of managers and supervisors scheduled.
- ❑ Feedback mechanism in place.
- ❑ Metrics to measure the progress and eventual success of the deployment identified and ready to use.
- ❑ Coaching and encouragement for managers, employees and stakeholders defined and ready — including those who will do the coaching.
- ❑ Support available; for example, facilitators, coaches, tech support, assistants, etc.
- ❑ Obstacle removal in place.
- ❑ Templates, checklists and other tools ready and available.
- ❑ Information on our web page done.
- ❑ Processes that will be impacted by the deployment identified and corrected.
- ❑ Changes to our quality manual completed.
- ❑ Manuals, guides and other reference materials that will be needed by the users ready and available.
- ❑ Executives briefed and ready to fully support the deployment.
- ❑ Human resources (HR) informed.
- ❑ Training department informed.

Deployment of a New Tool

Planning Document

This template may help you secure assistance from your executives. It is also a good point to start the process of deployment. It may even become a standard template for deployment of future tools and methodologies. Modify this template to meet your specific objectives within your organization.

1. Tool or methodology to be deployed.
2. Objective of the tool deployment.
   - What will be the outcome once the tool has been successfully deployed?
   - Why are you deploying this tool?
3. Who is the expert on this tool?
   - Who is the internal person who knows the tool and is willing to provide guidance to others?
4. What documentation is available?
   - Models.
   - Software.
   - Books.
5. Are there any external resources available?
   - Consultants.
   - Instructors.
   - Facilitators.
6. What are the risks involved in the deployment?
7. How will we pilot the tool?
   - Which team, organization or group will be the first to apply the tool to actual work.
8. What will be our communication plan?
   - With executives of the organization.
   - With employees.
   - With the pilot team.
   - With other stakeholders.
9. What will we measure to assure us that the deployment has been successful?

Deployment of a New Tool

Tips for Successful Deployment

There are a myriad of things you must consider before you deploy a new tool in your organization.

Brainstorm with your team to find innovative ways to ensure the success. Your corporate culture, industry and reporting structure will help you determine the best way to embed the new tool.

Key Questions

- How can we make the new tool a habit in our organization?
- How do we turn the novelty of the tool into the normal way we do things?

10 Tips for Successful Deployment
1. START SMALL — PICK A SIMPLE PROJECT FIRST *The intent at this stage is to ensure that your team knows how to use the new tool.*
2. CONSIDER THE CULTURE OF THE ORGANIZATION *If your organization has not used this tool before, you need to lead your team to understand and embrace it. Some will push back, and you need a plan to deal with some skepticism and denial.*
3. PROVIDE THE RIGHT TRAINING TO THE RIGHT PEOPLE AT THE RIGHT TIME *Your team, and even some managers, may need training. You must provide the skills necessary to succeed.*
4. ENSURE CROSS-FUNCTIONAL PARTICIPATION *Most tools provide maximum benefit when people from several functions participate. Moreover, most projects involve people from several areas of the organization, and they must be included.*
5. PUT IN PLACE A PROACTIVE FEEDBACK MECHANISM *Enable your team to provide you with input on the new tool.*
6. BE AVAILABLE TO REMOVE OBSTACLES *Inevitably, your team will run into obstacles when implementing a new tool for the first time. Put in place a mechanism that will enable you to remove these obstacles; otherwise, the team may reject the tool altogether.*
7. DEVELOP AN ESCALATION STRATEGY *Some of the issues and challenges that will emerge in deploying your new tool will need the attention of your managers and other executives. Plan an escalation process before things come to a boil.*
8. MAKE INFORMATION ABUNDANTLY AVAILABLE *When first deploying a tool, your team, managers and executives will need information. Make certain that information is abundant and conveniently available to all.*
9. MAKE COACHING AVAILABLE *Your team members will need one-on-one assistance. Even if you provide training, individuals will have questions and other barriers that you must help remove.*
10. CELEBRATE *Recognize progress — celebrate small and large victories. Everyone needs reassurance that progress is being made.*

Problem-Solving Process

NEW TOPIC

Problem Solving Process

Introduction

Program managers spend considerable time solving a stream of problems due to the fluid nature of technology. Many of these challenges are straightforward and can be readily dispatched.

However, sometimes tough problems are under-managed. Organizations wind up drifting, unsure of the reasons for change, the severity of the problem or the path forward.

A disciplined problem solving approach should be applied for the most difficult crises that will shape the direction of the business.

General Rules for Solving Problems

- Do not underestimate the need to apply a formal process to solving a problem. Some in the team might dismiss it as trivial or something that can easily be resolved.
- Beware of the easy solutions. Many times the problem is a symptom of a much larger issue. There is always the desire to jump to the solution without the process to diagnose the real root of the problem. Here is where a formal process of solving problems plays a very important role.
- Widely communicate the solution. Frequently a problem has been solved, but the majority of the team does not know it. The process for communicating the disposition of a problem should be as rigorous as the process to solve it.

Overview of the Problem Solving Process

1. Identify the problem. Come to a common agreement on the nature and extent of the issue.
2. Generate criteria for a solution. Select which business measures will guide the decision process.
3. Generate alternative solutions and test them against the business criteria you have selected.
4. Conduct diligent reviews on key options. Prune the options to the few that merit more detailed analysis against the criteria.
5. Decide! Make a clear decision and then back that decision with a detailed implementation plan.
6. Communicate the solution to all stakeholders.
7. Learn from the process

Important

TIPS is the acronym for Theory of Inventive Problem Solving, and TRIZ is the acronym for the same phrase in Russian.

TRIZ was first developed in 1946 by Genrich Altshuller and his colleagues in the former USSR and is now being developed and practiced throughout the world.

TRIZ research began with the hypothesis that there are universal principles of invention that are the basis for creative innovations that advance technology. Furthermore, if these principles could be identified and codified, they could be taught to people to make the process of invention more predictable. The research has proceeded in several stages over the last 50 years. More than two million patents have been examined, classified by their level of inventiveness and analyzed to look for principles of innovation. The three primary findings of this research are as follows:

Problems and solutions were repeated across industries and sciences.

Patterns of technical evolution were repeated across industries and sciences.

Innovations used scientific effects outside the field where they were developed.

For additional details visit: http://www.triz-journal.com.

Detailed Process: Problem Solving

1

Prepare for Your Problem Solving Meeting

Being well-prepared will make your meeting more successful and quicker:

- Gather all materials that will help you add clarity to the process; for example, technical documentation, reports and drawings.
- Select the attendees based on their ability to help solve the problem. Do not hesitate to invite people who are not directly involved with the problem but can add value to it. For example, experts on the topic and outside consultants.
- Ensure diversity of participants; the more points of view the better. For example, combine people who are involved in the problem with those who are not. People from other functions may also be helpful, as will some of your suppliers and vendors.
- Take care of all logistics; for example, reserve a venue that is large enough for your team to work.

2

Start Your Problem Solving Meeting

Share the perceptions of the problem: How does each team member see the problem? This is a critical step, as the term "problem" may mean different things to different people. Spend enough time in this definition phase — don't shortchange it.

Write the various perceptions on the flip chart. You can also use yellow stickies.

Clarify the definition of the problem by agreeing on a single wording of the problem.

Identify and document on a flip chart the expectations for a solution from your team members. As with the definition of the problem, the perception of a solution will vary. It is important to reconcile; i.e., find a common purpose.

Important

We have various ways to identify something that is not desirable; for example, we might call it a flaw, failure or fiasco.

Some of these terms carry a certain stigma or prejudice. Some cases imply punishment.

It is important that your team agrees on a term that is not politically loaded, but rather brings objectivity to your dialog. You cannot afford distractions.

3

Confirm Problem Statement

At this point you will have spent a considerable amount of time writing and re-writing the problem statement.

Write the problem statement one last time, as clear and readable as possible.

Validate with your team by asking everyone to agree or disagree with the statement. Does it truly represent the problem to be solved?

This is the most critical step of this process, the clear articulation of the problem, plus the support of the entire team.

You may need to spend additional time refining the problems statement, which would be time well-spent!

Do not proceed until there is clear alignment.

Important

The problem statement — the one you and your team have crafted — will most likely be the hardest step in the entire process.

Don't hesitate to rewrite the problem statement several times; it's worth the effort.

Spend all the time you need to ensure clarity and consensus among team members.

Do not proceed until everyone understands the statement of the problem.

Continued

For comments, questions or additional copies please e-mail the author: risk@rapidinnovation.com.

Problem-Solving Process

Cont'd

4 Profile the Playing Field

Analyze the problem based on the:

- **GIVENS:** Things that cannot be changed as part of the solution to the problem such as raising prices far in excess of what the customers are willing to pay.
- **FORCES AT WORK:** Those that are acting for and against the solution. Also known as Force Field Analysis.
- **MUST-HAVES:** Things that are non-negotiable or cannot be changed. Very similar to the givens, but with a focus on unacceptable trade-offs; for example, a reduction in product performance.
- **DESIRABLES:** Things that would be nice to have as part of your solution.

If appropriate, review the initial objectives of the project.

The outcome of this section is a single definition of the problem — the problem statement — and a set of criteria that bind the problem (a combination of steps 1 and 2).

5 Define and Document Solution Criteria

- Describe an effective solution (use the Yellow Sticky Protocol).
- Identify metrics to measure success of solution.
- Identify and list costs, customer needs and resources.
- Define the must-have criteria.

6 Generate Alternative Solutions

- These are possible solutions, not necessarily real ones. This is a critical step in order to tap the creativity of your team.
- The rules of brainstorming apply, such as don't criticism and keep an open mind. This is not the time to eliminate or stifle ideas.
- Use a brainstorming tool like an Affinity Diagram.

Important

We recommend using yellow stickies for the step of generating alternative solutions.

Ask your team to write down every possible solutions to the problem. Then ask your team to read each suggestion before posting every yellow sticky on the wall.

You can use the Affinity Process to organize and prioritize the yellow stickies.

7 Evaluate Alternatives Against the Criteria

- Evaluate the proposed solutions (step 3) against the developed criteria (step 2).
- Retest if you have other alternatives to consider.
- Don't underestimate the impact of varying skill levels and cultural gaps.
- Develop a matrix and consider combining alternatives.

Note: You may need to do research in order to validate that the alternatives are viable and achievable. This may take time, even require another meeting. Some teams opt to simply select the alternative, assign responsibility and proceed without validation.

See Also
Critical Tools

Continued

Problem-Solving Process

CONT'D

5

CONFIRM YOUR SELECTION OF SOLUTIONS

- Ensure clear understanding and alignment behind the preferred alternatives and the criteria.
- Conduct due diligence on the top options.
- Complete the fact-finding.
- Document the key assumptions.
- Document risks.
- Check for unintended impacts.

6

DEVELOP YOUR ACTION PLAN

- Develop a detailed plan to verify or validate the assumptions.
- Develop a plan to address the identified risks and uncertainty.
- Establish ground rules for the implementation of the solution. For example:
  - How are you going to deal with changes?
  - How will you monitor progress? Will you track the metrics, milestones or compliance?
  - When and how will you decide that the problem has been solved?
  - Who will be responsible for implementation?
  - Who will be responsible for closure?
  - Who, when and how will you declare victory?
- Close any gaps concerning resources, skills and culture (a gap is a misalignment between "what it is" and "what it needs to be"; for example, you need more resources that you currently have).
- Communicate the decision to all stakeholders.
- This step is all about ensuring execution; that is, the implementation of a road map that will guarantee the solution of the problem. This is a high-risk area, as many solutions fail to be implemented due to the lack of a plan of action.

LEARN AND IMPROVE (KAIZEN)

Your risk management process will improve significantly if you constantly add the lessons you've learned.

- Debrief the team on the process used to solve the problem.
- Identify areas of improvement.

Problem-Solving Process

Problem Solving

Flexibility Matrix for the Problem Solving Method

Strengths	Weaknesses
A traceable path of problem to solution.	Need to watch for hidden boundary conditions.
Ensures a lot of input from everyone.	A solution will be found, but it may be politically incorrect.
Participative; creates ownership and encourages diverse solutions.	It involves hours of work, and it may be tedious.
Appropriate for difficult problems.	
A way to dispose of a problem once and for all.	
A disciplined way to identify the real causes of a problem, not just opinions and anecdotal information.	

Flexibility

- You can vary the size of the team depending on the complexity of the problem.
- You can create additional steps to suit the culture of your organization.
- You can do the process in two or three sessions to avoid fatigue.
- Include the documentation in the archive of your project in case the problem shows up again.

The Problems that Refuse to Die

With some regularity, there are problems and issues that have been solved, yet they persist and continue to distract the organization. This may be due to a lack of proper communication; specifically, the problem was solved but only a few in the organization know it. This can be addressed by a proactive, high-energy communication effort. For example, clearly announce the solution during review meetings and send notices out specifically on the conclusion of the issue. It helps to be emphatic.

The other reason for this type of problem is that some in the organization do not wish it to go away. Keeping the problem alive is one way to fight the chosen solution. One way to deal with this difficulty is to use the formal process outlined in this document, which allows you to cover all the possibilities and overtly deal with and dispose of the problem — this is called an intervention. Simply put, deal with the issue directly.

Other Tools for Solving Problems

There are several tools that provide a fine framework for solving a problem; for example, the Affinity Diagram, the Fishbone Diagram (Ishikawa Diagram), Karnaugh Maps (2X2 matrixes) and various cause-and-effect diagrams. Some of these are covered in this document, and the rest are readily available on the Web and in other handbooks and guides.

More advanced methodologies are also available to tackle very complex or esoteric problems; for example, FMEA and TRIZ. These methodologies are also documented on the Web.

Avail yourself of these tools. We suggest that you become proficient in at least one or two.

Karnaugh Maps: 2x2 Matrixes

NEW TOPIC

Prioritize Items from a Selection of Many

Introduction

Karnaugh Maps, also known as Karnaugh Diagrams, have their origins in mathematics and are related to Venn Diagrams . They are especially useful for generating the most optimal Boolean expression. By varying the criteria and the values, you can apply this tool to select or prioritize many choices.

A Karnaugh Map is a 2x2 matrix. But it is also a participative tool that enables a group to prioritize or select items from a group of many, using a set of criteria. It is a simplification of a FMEA. In short, it enables you to make better and faster decisions.

Applications

Facilitators are often faced with helping a team make a decision or a selection to set priorities from an array of many options. For example, they must determine the risk factors of a new deployment, select key features on a new product, determine the importance of tasks or projects and much more.

The scientific community has been successfully dealing with this issue for some time. Over the years, engineers and scientists have developed tools and methods to help them cope with unanticipated failures. They have learned successful ways to anticipate what may go wrong in order to prevent it. From risk analyses to FMEA (Failure Model and Effect Analysis) to design of experiments, these methods, when well-understood and carefully applied, have proved to work.

Generally, when choosing we have two criteria to use — for example, "urgent" and "important." While some things can be very important, they are not necessarily urgent. Thus, the ability of a team to separate all tasks based on the set of criteria will result in a better deployment of limited resources.

Another example is risk selection. Take any project and you can ask your team to identify those things that are likely to go wrong, and the negative impact on the project if they do. These two criteria will help the team focus on those things that are likely to happen, and will cause serious damage to their project. Thus, rather than focusing on all the things that can go wrong, they can focus on the vital few.

Why 2x2 Matrixes

- The 2x2 matrix is attractive because it is graphical and simple and enables group participation.
- It engages everyone on the team, and allows for innovation and breakthrough.
- It can be as rigorous as you need it to be — i.e., the results can be made statistically defensible.
- It is very flexible, from prioritizing to selecting, choosing and more.

General Rules for Karnaugh Maps

- You should first be familiar with the concepts of Affinity Diagrams, or KJ (Kawakita Jiro) method.
- A session to develop a successful 2x2 map will take three to four hours.
- The challenge is to select the right criteria for each of the axes. The Y axis can represent one criterion such as urgency. You can assign values; i.e., a scale, such as 0 to 10.

Definition

Facilitator

Anyone who is leading a team to success; this includes project managers, team leaders, program managers, supervisors, leads, etc.

Pair-Wise

A process of comparing entities in pairs to determine preference.

The method is used in statistical analysis, specifically probability.

L.L. Thurstone first introduced this approach in 1927, using pair-wise comparisons for measurement.

For our purposes, a group or team must express a preference such as priority between two mutually distinct alternatives in order to meet an objective.

Karnaugh Maps: 2x2 Matrixes

- The X axis (for example, important) is your second criteria. Likewise, you can assign to it a value.
- Once you have a way to plot items based on the criteria (e.g., importance and urgency) place yellow stickies in the appropriate quadrant. For example, the items in the upper-right corner are more important and more urgent than the others.
- The color of the yellow stickies and their size can also represent additional criteria such as cost and difficulty of implementation. We recommend that you start with just two variables.
- We suggest that you limit your Karnaugh Maps to no more than three variables. Because Karnaugh Maps are a visual tool, it is best to limit the number of variables so as not to confuse the team.

Multiple Variables

Can you use more than three variables? Yes, you can. You can use a weighted-average method through the use of an Excel spreadsheet. You can have as many variables as you wish. But, it is important to show the results in only three variables. There is the tendency to complicate matters beyond what you need. Be careful! Let your objectives determine the number of variables. In our experience, most instances only require two or, at most, three.

The Scales

Generally, a scale of 1 to 10 on both axes is sufficient. It is important that the team agrees with the definition of the scale; i.e., decide what each incremental level means. Some teams prefer to use a 1 to 5 scale, which is also acceptable.

Remember, this approach is **QUALITATIVE**. A **QUANTITATIVE** scale is also acceptable, but we highly recommend that you choose these sparingly. The additional accuracy provided by a quantitative scale, in most cases, is overkill. Use quantitative scales when the topic at hand is critical and of extreme importance to your team or company.

A Better Way to Prioritize Affinity Diagrams

You have an alternative to help you prioritize your Affinity Diagram: You can choose the method outlined in the Affinity section (Prioritizing an Affinity Diagram) or you can choose the Karnaugh Map (KM) approach. Both are valuable and useful, although Karnaugh Maps provide more focus and perhaps a bit more accuracy. Karnaugh Maps also provide a more rational and organized approach than Affinity Diagrams. With Affinity Diagrams, you prioritize using a single variable, but Karnaugh Maps allow for two or more variables. For example, you could set the x and y axes to indicate "cost to develop" and "importance to the customer," and these two variables, called pair-wise, combine more than one criterion. You can even add a third variable; for example, the color of the stickies denotes "difficulty to develop." We suggest that you limit your Karnaugh Maps to no more than three variables. For example, the x axis is "cost to develop," the y axis is "importance to the customer" and the color of the Stickies is "difficulty to develop." In short, because Karnaugh Maps are a visual tool, it is best to limit the number of variable so as not to confuse the team. For examples of other combinations of variables, see examples of pair-wise criteria and their application later in this section.

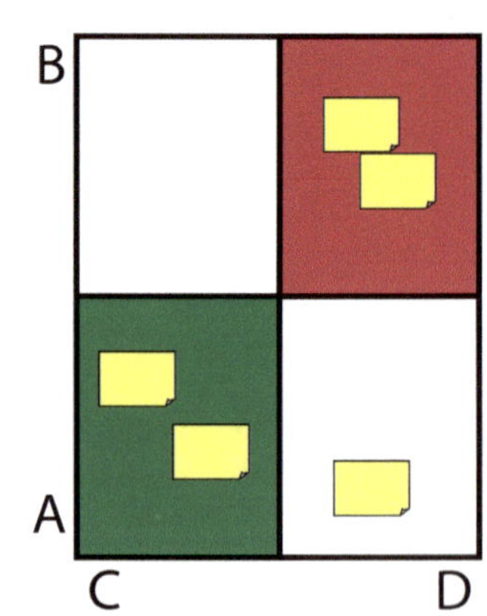

Typical configuration of a 2x2 matrix (Karnaugh Map). Notice the four quadrants.

Breadth of Uses for Karnaugh Maps

Once you learn to use them, you will find that they can be used in areas well beyond customer contacts. The famous 2x2 matrixes have been around for many years. They can be used to prioritize or make decisions. In short, any time a decision needs to be made, you can use Karnaugh Maps.

Karnaugh Maps: 2x2 Matrixes

Example: Karnaugh Map (2x2 Matrix)

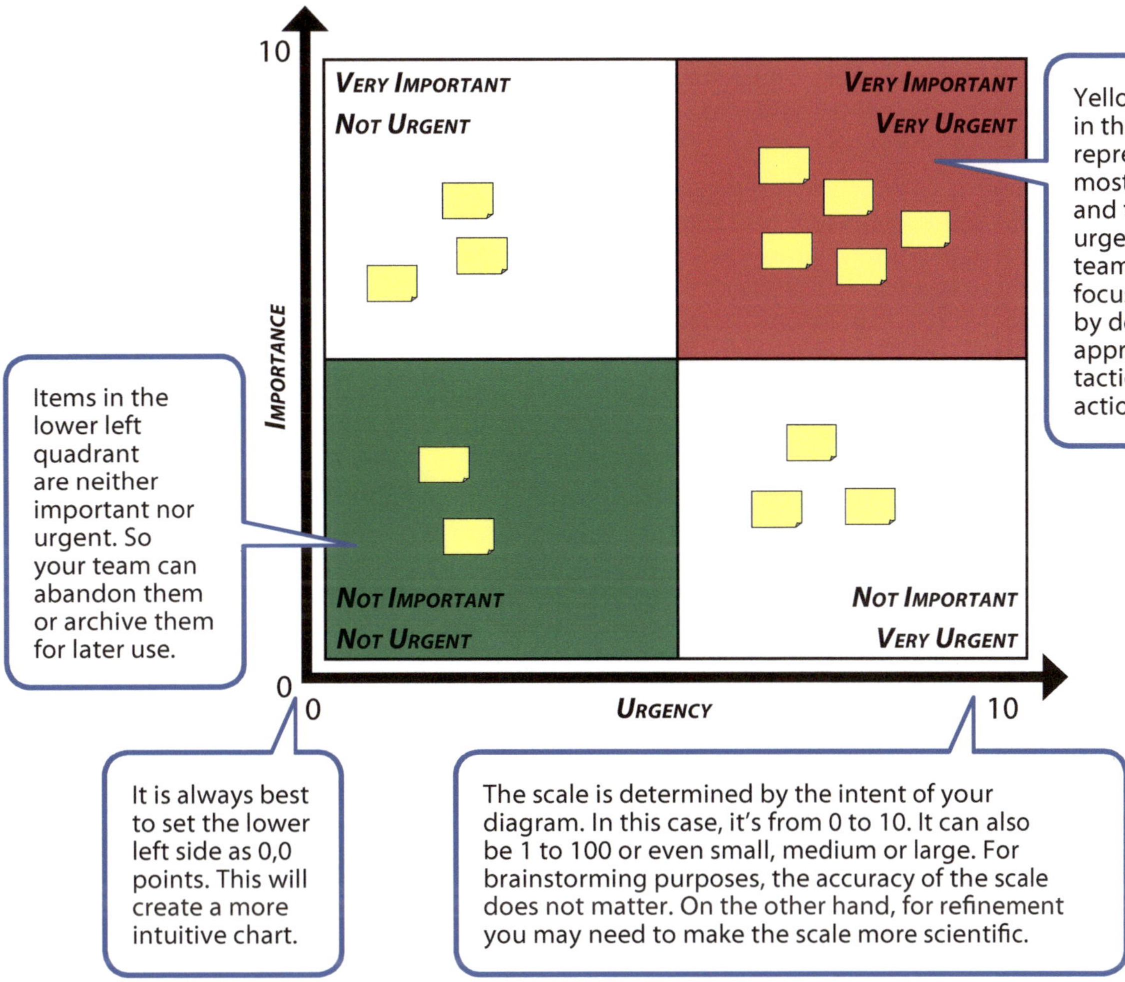

Karnaugh Maps: 2x2 Matrixes

Pair-Wise Comparison

Pair-wise comparison generally refers to any process of comparing entities in pairs to judge which of each pair is preferred or has a greater amount of some quantitative property. Pair-wise comparisons are relational — unlike Cartesian X and Y coordinates, where one is the function of the other. The method of pair-wise comparison is used in the scientific study of preferences, attitudes, voting systems, social choice, public choice and multi-agent AI systems. In statistical analysis, it's used to determine probability.

Prominent psychometrician L. L. Thurstone first introduced a scientific approach to using pair-wise comparisons for measurement in 1927, which he referred to as the Law of Comparative Judgment. Thurstone linked this approach to psychophysical theory developed by Ernst Heinrich Weber and Gustav Fechner. In psychology literature, it is often referred to as Paired Comparison.

Application to Product Development

In product development, your team can use pair-wise comparison to express a priority between two mutually distinct alternatives in order to meet an objective.

Many pair-wise criteria are possible. We offer a short list of examples to help you and your team develop criteria that fit your needs exactly.

1. **CHARACTERIZING RISK:** The criteria of "Likelihood of Happening (Probability)" and "Severity if it Does Happen (Impact)" — often expressed as a P-I Factor — maps risks that are most likely to occur and will have the most severity if they do. You can use the results to manage risk.
2. **SELECTION OF OPTIONS:** For example, when selecting projects or courses, use "Attractiveness" and "Cost" to determine best value.
3. **MAKING CHOICES:** When making trade-offs, try "Ease of Use" and "Cost of Implementation." "Achievability" and "Benefits" are simple criteria you could develop to have real meaning — for example "Achievable within the Fiscal Year" and "Financially Beneficial."
4. **DEFINING PRIORITY:** Use "Urgent" and "Important" to prioritize the allocation of resources.
5. **DOING A GAP ANALYSIS:** Use "Current Level versus Desired Level" and "Difficulty in Achieving."
6. **DEFINING COMPLEXITY:** Use "Complexity" and "Difficulty in Achieving."
7. **DOING RESOURCE ALLOCATION:** Compare "People" or "Resources" on the Y-axis with "Priority of Action" on the X-axis. The priority can be defined in strategic terms or some type of urgency scale.
8. **DEFINING THE COMPETITION:** Map the need to defend a segment, abandon it or create a new segment. One example: "Level of Competition" and "Urgency to Defend." Another example: "Level of Competition" and "Critical Market Segment."
9. **ENSURING FIT:** Use "Fit with Business Objectives" and "Cost of Implementation."

Karnaugh Maps: 2x2 Matrixes

Detailed Process: Prioritizing a Karnaugh Map

1

Set Objectives for Your Karnaugh Map

For example:

- Determine risks and rewards.
- Prioritize from a choice of many.
- Select options based on certain criteria.
- Choose a course of action, given certain criteria.

Important It's a good idea to prepare a blank 2x2 matrix in advance. Tape two or more sheets of flip chart paper together and use markers to draw the matrix. A white board is an alternative, but upon completion you need to carefully gather the yellow stickies.

Important After a brainstorming session, you may end up with as many as 300 yellow stickies. Scrubbing is the step to eliminate duplicates, redundancies and any other non-value-added yellow stickies.

Your goal is to end up with approximately 30 or fewer yellow stickies for the last step. On the other hand, if you end up with many more, and you feel they are important, then proceed.

2

Process the Information

Once all the yellow stickies have been collected, start scrubbing the information to ensure that every yellow sticky communicates a clear, distinct idea.

This step can take 30 to 45 minutes.

If you have too many yellow stickies, then you need to affinitize (group) them into main categories.

Important The selection of the two criteria for the X and Y axes is critical. To help determine the criteria, spend sufficient time with the team discussing "why" or "how" the final information will be used.

3

Define the Criteria and Scale

Select the pair-wise criteria for the X and Y axes. For example:

- Important vs. Urgent
- Expense vs. Ease of Implementation
- Risk vs. Rewards
- Important to Customers vs. Expense to Develop

Set your scale. For example:

- Small, medium and high
- 1 to 10 or 1 to 100
- A, B and C

Important You can define a third criterion via the color or size of the stickies. Use this option sparingly. Furthermore, when first deploying the Karnaugh Map tool, we recommend you begin with just two criteria.

Continued

Karnaugh Maps: 2x2 Matrixes

Cont'd

See Also **Pair-Wise Criteria**

4

Plot the Information on the Matrix

Using your criteria, place each yellow sticky inside the 2x2 matrix. Do this in collaboration with your team.

You can use several methods of voting to assign value to each of the yellow stickies — for example, count hands, use dots, assign numerical values to later average. The more critical the topic, the more formal the vote should be.

Don't dispose of the plot, as you may need it for future reference.

5

Reach Your Conclusions

Placing all the yellow stickies does not complete the process!

The value of your 2x2 matrix is in the decisions that you make, the action-items that are generated and the follow through to act on these directives — simply put, the process is not over until something happens.

6

Develop Your Action Plan

Remember, it's not an action plan unless it clearly identifies "do what," "by whom" and "by when."

Document your action plan.

Follow up until all actions have been completed.

NEW TOPIC

The 5 Whys

Help Customers Reflect Their True Needs

This is a simple methodology to obtain valuable information from customers. As the name implies, in the literal sense, it means asking the customers "why" five times in order to get to the root message — i.e., the most fundamental message concerning the needs of the customer. This methodology originated in Japan and is typically associated with the process of identifying why failures occur. It is part of Root Cause Analysis.

In the context of customer interviews, the 5 Whys serves the purpose of getting past the stock answer. It allows you to peel away the layers of information until you get the valuable information underlying the needs of the customer. The 5 Whys is an excellent tool to get your customers to tell you their unarticulated or unstated needs. The process of asking why causes the customers to reflect on their actual needs.

General Rules for the 5 Whys

1. You can use as many whys as needed. While the name of the methodology is "5 Whys," it simply means that you can continue to ask why until you get useful information. In some cases, three whys will suffice, yet in other cases as many as seven will be needed. Use your judgment, and stop asking when you feel you have obtained the root message or the valuable information you are seeking.
2. Avoid the toddler's curse, when a 3-year-old child drives you crazy asking "why." Instead, let the customer talk, then ask why. You can also use other ways to ask — for example: "Tell me more" and "Why is that?"
3. Help the customer by restating her previous answers or by repeating the questions. For example: "You said that this product is difficult to use, and that this is because the menu system is not intuitive; now, can you describe why the menu system is not intuitive?
4. The intent of the 5 Whys is to get to a level where you get actionable input and not generalities. Actionable input is customer information that you can actually use in developing requirements and features.
5. The 5 Whys is not intended to be used as an interrogation tool. That is, one does not actually ask "why" five times. We encourage you to use your own style to apply the concept.
6. Your role is to enable your customer to talk and express his "root message."

The 5 Whys

Examples: The 5 Whys

Example 1: The Main Pump Failed

Q

Q: Why did the main pump fail?

A: The coupling from the main intake to the pump broke and created a shutdown of the line.

1st Why

Q: Why did the coupling from the main intake fail?

A: The "O" ring failed under pressure.

2nd Why

Q: Why did the "O" ring fail under pressure?

A: The night temperature fell to -20 degrees and the "O" ring was not specified for low temperatures.

3rd Why

Q: Why was the "O" ring not specified for low temperatures?

A: No one expected these pumps to be sold in Alaska.

A

ROOT CAUSE: Insufficient environmental specs to enable safe operation of the pump under arctic conditions.

CORRECTIVE ACTION: Review all environmental specs and redesign for arctic conditions.

Example 2: The Machine Operator Received an Electric Shock when Using the Pick and Place Machine

Q

Q: Why did the machine operator received an electric shock when using the Pick and Place machine?

A: The operator touched the high voltage line.

1st Why

Q: Why did the operator touch the high voltage line resulting in shock?

A: The safety cover to the high voltage lines was open and the operator reached inside.

2nd Why

Q: Why did the operator reached inside the cover?

A: The operator thought that there were adjustment controls inside the cover.

3rd Why

Q: Why did the operator think that there were adjustments inside the box?

A: The box had no label and was opened.

A

ROOT CAUSE: Lack of proper safety marking and also a safety box that has no lock nor "Poka Yoke."

CORRECTIVE ACTION: Place warning labels prominently on and about the safety box. Add a lock that only the maintenance technicians can use. Add an automatic disconnect when the box is opened.

The following examples illustrate the concept of the 5 Whys. In both cases there are only three follow-up questions. In some cases, it could take seven or more follow-up questions. The 5 Whys concept can be used in many areas of inquiry.

Yellow Sticky Protocol

Effective Group Participation

Benefits

The ability to ensure that everyone in a group participates is greatly enhanced through the use of Post-It Notes® (trademark of 3M), known in the vernacular as "yellow stickies." Another benefit of yellow stickies is that they provide a self-documenting process; rather than writing things on a flip chart, which is time-consuming, yellow stickies significantly accelerate the documentation. Finally, yellow stickies are portable; that is, one can move them, hand them over, rewrite them, etc.

Some individuals, particularly senior executives, are not fond of yellow stickies. This is due to a misunderstanding as to who uses them. But this is a serious tool that will produce better results, and more quickly. These people need to be reassured that, barring a technological breakthrough, yellow stickies are still a fine tool that everyone, including executives, can use.

The process favored in this guidebook relies heavily on the proper use of yellow stickies.

Essential

"The process favored in this guidebook relies heavily on the proper use of yellow stickies."

Establish the Protocol

To derive the benefits of using yellow stickies, the facilitator must establish a protocol — an articulated set of guidelines as to the way the information will be presented in each yellow sticky.

The protocol is a way to create a safe environment. For example, use the protocol to assuage members who may have reservations about their poor spelling, grammar or handwriting.

The protocol is a way to choreograph the process so it yields maximum benefits.

For the sake of productivity, let alone sanity, the protocol must be enforced. All team members must participate and follow the protocol. Every yellow sticky must conform to a standard to be useful.

There is flexibility, so you need not do it exactly as in the corresponding detailed process. The overall goal is to be able to track and process a large amount of information.

Your team must agree on the way the yellow stickies will be structured. Once you have reviewed and highlighted all of your notes, you proceed to transfer the information to the yellow stickies. You need a protocol as to the type of information and its location. This is critical to provide continuity, traceability and consistency

Plan Ahead

Schedule a three to five hour meeting for a team of up to 15 people in a large conference room. Prepare topics and objectives for the meeting as well as the introduction. Gather materials:

- Post-It Notes® (3M brand, model #655 3x5).
- Sharpie® (Sanford brand, fine point, black markers) for all attendees.
- Flip chart paper and easels — two are best.

Yellow Sticky Protocol

Level 3 Detailed Process: Yellow Sticky Protocol

1

PREPARE YOUR MATERIALS

- 3x5, 3M-brand, Post-It Notes® (#655) — One pad per participant. Consider that several of the participants will have a sizeable number of yellow stickies — in some cases as many as several hundred.
- Sanford-brand Sharpie® markers, fine point, black.
- Flip charts or a place to hang the yellow stickies once filled.

Important: You also will need the participation of everyone who made customer visits.

2

MAKE THE ANNOUNCEMENT IN FRONT OF THE GROUP

Face the audience, make eye contact and say...

"We are going to use yellow stickies in this meeting.

There are three key reasons why this is important:

- It is a self-documenting process, so it saves time and allows for documentation after the meeting.
- It allows everyone to participate.
- We can move the yellow stickies around depending on our needs — no need to rewrite, just move them.

Important: Caution: Some individuals, particularly senior executives, are not fond of yellow stickies. This is due to a misunderstanding as to who uses them. But this is a serious tool that will produce better results, and more quickly. These people need to be reassured that yellow stickies are a fine tool that everyone can use.

3

EXPLAIN THE PROTOCOL TO THE GROUP

"In order to take full advantage of the tool, we must follow a simple set of rules..."

For example:

1. Don't worry about spelling or grammar; no one will check it. This goes for everyone in this room.
2. Don't write on the sticky side of the note.
3. Write landscape (horizontally) oriented.
4. Use your Sharpie® marker. This will force you to write larger than usual and more legibly, and it will enable others to read notes from a distance.
5. If you make a mistake or change your mind, simply toss it and start a new one.
6. Write only one item per yellow sticky. Only one question, one idea, one quote.

Important: Encourage everyone to write legibly. Ask them to print. Many of us do not have the best of penmanship. Explain to the participants that they are writing for others to read! It will save time and reduce confusion.

4

START THE PROCESS

See Also: Affinity Diagrams

For comments, questions or additional copies please e-mail the author: risk@rapidinnovation.com.

Yellow Sticky Protocol

Example: Yellow Sticky Protocol

The topic of conversation from your discussion guide.

This must match the topics contained in the discussion guide that was created for this set of customer interviews.

An agreed-to indication of the importance, relevance or some other priority. In this example, "H" indicates "High" passion or importance of the statement from the source, whereas "M" would indicate "Medium" and "L" would indicate "Low."

Your team can develop similar priorities, depending on the objectives of your visits. Having a priority set will save considerable time during the later stages of this process. Keep it simple; otherwise, your team members will be confused and misinterpret the code.

This area is reserved for the message from the interviewee — the voice of the customer.

It can be a quotation, a fact, a finding, information or any sentence that is relevant. This is extracted from the notes you took during the interview — you must be able to trace it to your notes.

The Protocol

Topic   H   Customer's title

"Actual quote from the customer, extracted from the interview notes."

Your name/initials   Page, Interview #

The upper right corner is reserved for the title or function of the person who was interviewed.

You need to develop codes to be used here to help you understand the context, particularly if your team interviewed customers from different functional groups.

The name of the customer is not relevant.

Quote marks (" ") are reserved for actual customer quotes! This is a critical rule.

Quotes are always verbatim messages from the customer. Do not use quotes if the customer did not actually say it.

A simple code will enable you to trace the information to your notes, and thus to an actual interview.

It is difficult to remember who said what. During the processing and various meetings, you will want to refer to your notes to clarify the context of the statement.

This is simply a tracking mechanism. You should be able to track back to the actual source of the information.

Example

Ease of use   H   Design engineer

"More than three layers of menus make products difficult to use."

Fred T   Page 1, 2nd interview

Yellow stickies are landscape (horizontally) oriented)

This example shows a super set of the Yellow Sticky Protocol. You need not use all of the items, just those that will enable you to track and code the critical information. When you have several hundred yellow stickies, you will rely heavily on the protocol to make sense of it all.

Affinity Diagrams

NEW TOPIC

Organize Information and Reach Consensus

Introduction

Affinity Diagrams are a very practical, easy to use method that can be applied in many areas once you learn how to facilitate the process — you lead your team through the process, and your team members look to you for instructions and guidance to reach the results.

Affinity Diagrams can be used in many areas of team work:

- Prioritize topics from a list of many
- Collaborate to solve a problem
- Develop a strategy
- Define group options and make choices
- Identify alternative paths or options
- Brainstorm discuss a set of issue

In short, Affinity Diagrams allow you to tap the power of a team by enabling everyone to participate and ensuring that everyone has the opportunity to innovate and create.

Definition

Affinity

A natural liking for or inclination toward somebody or something or a feeling of identification with somebody or something.

Definition

Affinity Diagrams

A tool to organize information and reach consensus."

Essential

"Affinity Diagrams are essential to the process favored by this guidebook."

Historical Background

Affinity Diagrams belong to a set of tools called the KJ method, which is named after Kawakita Jiro, the Japanese anthropologist who first introduced the approach in 1967. The approach has its roots in the values of Zen Buddhism.

Mr. Kawakita developed the method as a means of organizing diverse observations and qualitative information into useful, documented facts. He observed that humans are not linear thinkers; that is, ideas flow in a non-linear fashion. Further, humans tend to jump around when brainstorming. Thus, trying to structure brainstorming sub-optimizes the creative flow.

Why Affinity Diagrams

- They are simple and non-threatening. As the saying goes, "They're not rocket science." This creates an inviting environment for full participation.
- They are inclusive. Properly used, an Affinity Diagram allows everyone on a team to fully participate, regardless of stature or pecking order.
- They are a fine way to ensure consensus in a group, as well as to build the sense of team.
- They are a practical way of organizing seemingly unrelated facts.
- They are very conducive to the creation of clarity and communication within the team.

General Recommendations for Affinity Diagramming

1. Planning your session is essential to success. Please do not treat an Affinity Diagram process as a casual affair. Consult the process map in this document for additional information.
2. Select a room large enough to allow the team to move around. You may have to move furniture to create a clean space to work.

Affinity Diagrams

3. Affinity Diagrams work best when done by a group of about 10 people. Larger groups tend to fracture, meaning that some tend to marginalize themselves and not participate. Thus, groups larger than 10 require considerable facilitation and constant focus to ensure productivity.
4. Silence or talking? The general rules indicate the whole process should be done in silence to allow creativity to flow and avoid any one individual influencing the group. We believe that, in certain circumstances, talking is desirable. When you have a small group, say about three to four people, or when everyone is a peer and collaborative by nature, these are examples of when talking might help. Use your judgment and decide based on the dynamics of your group.
5. Use the traditional rules of brainstorming — for example, do not allow criticism of ideas, encourage participation by all team members and generate the largest possible number of ideas, including duplicates and repetition.
6. Remember the follow-up. Many fine sessions end up delivering little or no value due to the lack of action after the session. Ensure before you start that you are committed to following up well after the session is completed. Every session must have an action plan ("road map") to ensure that you produce tangible results.
7. It is best to use flip chart paper. We also recommend other types of large-format paper such as butcher paper, newsprint rolls and _. Do not hesitate to create a large space with paper. Remember that your team will be working in this area — six to nine horizontal linear feet of paper is best.
8. Avoid cryptic descriptions, such as one-word statements like "energy" or "easy." Encourage team members to use the proactive or active tense — for example, "There is not enough energy" or "Make it easy to use." Therefore, statements should have a noun and a verb to avoid ambiguity and enhance clarity.
9. The number of categories or columns should be kept to about five; more than that confuses the teams.
10. Create the feeding frenzy. It is fun and productive to encourage the team to generate as many ideas as possible. Set challenges — for example, ask everyone to generate five ideas. Do many rounds, create contests and give a prize to the person who generates the most ideas.

High-Level Process

1 — Start

Deliver instructions.

2 — Brainstorm

Participants create yellow stickies.

3 — Post

Place yellow stickies on the wall.

4 — Organize

Participants organize yellow stickies into categories.

5 — Prioritize

Participants prioritize categories.

6 — Develop Action

Assign responsibilities and develop a timeline.

Important! This is a high-level view of the process for an Affinity Diagram. It will help you navigate, without the complexity of all the details; i.e., the Detailed Process Map.

Affinity Diagrams

Detailed Process: Affinity Diagrams

1 Invite the Participants

Who should attend?

- All stakeholders in the process or project.
- Customers, if appropriate.
- Try to keep the group to 10 participants or fewer.

2 Prepare

- A large conference room
- Post-It Notes® (3M brand, model #655 3x5)
- Sharpie® (Sanford brand, fine point, black markers) for all attendees
- Flip chart paper and easels — two are best

See Also Yellow Sticky Protocol

3 Set Up the Topic

- State the topic in the form of a question. Ensure that everyone understands it. You may want to discuss it with the group.
- Reassure everyone that there are no wrong answers.
- Reassure everyone that grammar and spelling are not important.
- Explain how to use the yellow stickies — one per subject, use a Sharpie® pen, landscape (horizontally) oriented.
- No criticism.

4 Start the Process

- Give the question or topic, then ask everyone to write at least three yellow stickies and wait.
- Ask each person to read his yellow stickies as loudly as possible.
- Tell everyone to listen intently as others read their yellow stickies.
- Encourage participants to write additional yellow stickies as others present theirs.
- Collect all yellow stickies and place them on the flip chart or board at random.
- Continue until the group runs out of ideas or repetition sets in.

5 Scrubbing the Affinity Diagram

Clean up your Affinity Diagram once all the yellow stickies are written and placed on the board.

- Eliminate yellow stickies that are obviously irrelevant or trivial.
- Eliminate obvious duplications.
- Rewrite yellow stickies that are difficult to read.
- Rewrite yellow stickies that are not clear.

Important! For any and all changes you make when scrubbing the Affinity Diagram, make them with the agreement of the authors.

Continued

Affinity Diagrams

Cont'd

6

Start the Affinity Process

Rally the group to move up around the flip charts with the yellow stickies.

Ask participants to observe how the various yellow stickies cluster or group into subjects.

Ask the participants to start moving the yellow stickies, using their own criteria to cluster them, regardless of how others feel about it.

Do this in silence! No one can talk during this step!

Stop when there is little or no movement of yellow stickies.

7

Pick 'n Choose

Review each column or cluster that was created in the previous step, with participation by the group.

Ask the group to explain why the cluster or column exists. "Why are these yellow stickies together?"

Move the yellow stickies that do not belong in the column to a different one or a separate place on the flip chart.

Eliminate duplicates by choosing one that represents the topic and placing the others underneath.

For each column, assign a header that captures the topic. Write the header on a sticky notes, preferably a color other than yellow.

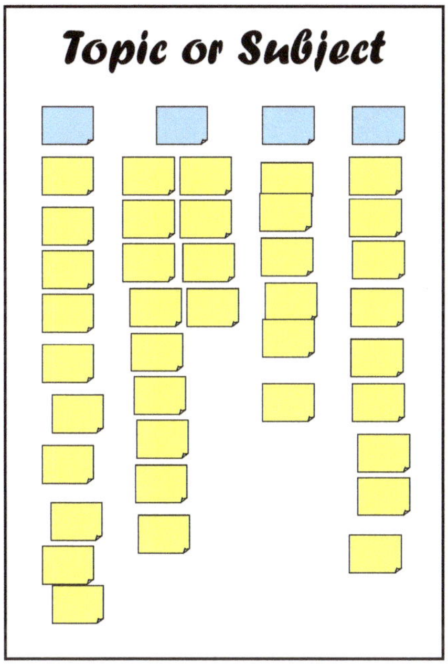

Yellow Sticky Approach

8

Prioritize

Once the notes are prioritized, you can begin to assign responsibilities and action items as appropriate.

Also, develop a timeline. This is a critical step!

Save the flip charts for archival purposes or to be transcribed if necessary.

Debrief the group on the effectiveness of the tool and ways to improve the process.

See Also Prioritizing Affinity Diagrams

See Also Karnaugh Maps

Prioritizing Affinity Diagrams

NEW TOPIC

The Process to Prioritize

Most often, Affinity Diagrams are used to prioritize a set of ideas, choices, things, etc. The criteria for prioritizing are determined by the initial objective that you set for the Affinity Diagram — for example, you might want to do an Affinity Diagram to assign responsibilities or action items. You might also want to pick a few from a selection of many. In short, use the same objective you set for the Affinity Diagram to select the method you will use to prioritize.

Group Consensus

Prioritizing is all about making decisions; keep this in mind as you work with your team. You want to reach consensus. 100 percent agreement is desirable, but not essential. Most of your prioritizing efforts will not have 100 percent consensus. That being said, prioritizing an Affinity Diagram is a team effort.

Select The Criteria

The key to prioritizing is the selection of the criteria used to make the decisions. Take your time, discuss with your team, and above all ensure absolute clarity. Criteria should have two or three elements; in a team environment and without computers it's very difficult to keep track of more than three criteria.

This process should be driven by the strategic importance of the project. For small, simple projects the criteria might be somewhat subjective. If the project is critical, then the criteria must be defensible — objective, verifiable and in full alignment with the strategic goals of your project. This is a qualitative exercise, so avoid criteria that are too complex.

Common Methods

1. ABOVE OR BELOW THE LINE

Draw a horizontal line at the midway, and ask the team to decide for each yellow sticky if it should be above or below the line. This is the quickest approach and one that most people find intuitive. You can prioritize the yellow stickies or the headers depending on your objectives.

2. 1 ROUND OF DOTS

This is a popular method because it's simple and familiar. Give everybody five to 10 self-stick colored dots to place on the yellow stickies based on the predetermined criteria. They can place their dots anywhere, as many dots as they wish per item. When done, simply count the dots to establish priority.

3. 2 ROUNDS OF DOTS

A variation on this theme is to do two rounds of dot placing with five dots for both rounds:

Round 1: Place five dots anywhere, with no guidelines.

Round 2: Place five dots, but only on two yellow stickies.

4. 3 ROUNDS OF DOTS

A more sophisticated method involves giving each team member 30 dots to be placed during three rounds:

Round 1: Place 10 dots anywhere, with no guidelines.

Round 2: Place 10 dots, but no more than two dots on any one yellow sticky.

Round 3: Place the remaining 10 dots, but only on two yellow stickies (five and five).

5. YOUR INVENTION

Invent your own system for prioritizing. As long as it's rational and leads to clear decisions, you'll be fine.

Level 3 Detailed Process: Prioritizing Affinity Diagrams

1

Select the Criteria

The set of criteria is generally two to three elements that will determine priority. The criteria will drive the decision process.

Take your time, discuss with your team, and above all ensure absolute clarity.

The simplicity or complexity of your criteria depends on the strategic importance of the project. As a rule, keep it simple.

Once agreement has been reached, write the criteria on a flip chart to help the team stay consistent.

Important! This process should be driven by the strategic importance of the project. For small, simple projects the criteria might be somewhat subjective.

If the project is critical, then the criteria must be defensible — objective, verifiable and in full alignment with the strategic goals of your project.

2

Start the Process to Prioritize

Q. Do we prioritize the headers, or each yellow sticky?

A. It depends on your objectives:

The preferred method is to prioritize the individual yellow stickies. Based on which categories the prioritized yellow stickies are in, you can infer the priority for the categories.

For example, if you use one of the Dots methods, count the number of dots in each category, and write it on the header. This will create an overall priority ranking.

Another approach is to ask the participants to place the dots on the actual headers — not the yellow stickies. This is faster, but it may miss an important message: Which yellow sticky is most important?

Use your judgment as to which voting method will serve you best.

Important! Prioritizing is a team effort. 100 percent agreement is desirable, but neither essential nor expected. The goal is consensus.

3

Prioritize

Choose from one of these common methods or invent your own:

- **ABOVE OR BELOW THE LINE:** Draw a horizontal line, and ask the team to decide for each yellow sticky if it should be above or below the line.
- **1 ROUND OF DOTS:** Give everybody five to 10 self-stick colored dots to place on the yellow stickies based on the criteria. They can place their dots anywhere, as many dots as they wish per item. When done, simply count the dots to establish priority.
- **2 ROUNDS OF DOTS:** A variation on this theme is to do two rounds of dot placing with five dots for both rounds:
  - Round 1: Place five dots anywhere, with no guidelines.
  - Round 2: Place five dots, but only on two yellow stickies.
- **3 ROUNDS OF DOTS:** A more sophisticated method involves giving each team member 30 dots to be placed during three rounds:
  - Round 1: Place 10 dots anywhere, with no guidelines.
  - Round 2: Place 10 dots, but no more than two dots on any one yellow sticky.
  - Round 3: Place the remaining 10 dots, but only on two yellow stickies (five and five)

Prioritizing Affinity Diagrams

Example: Affinity Diagram

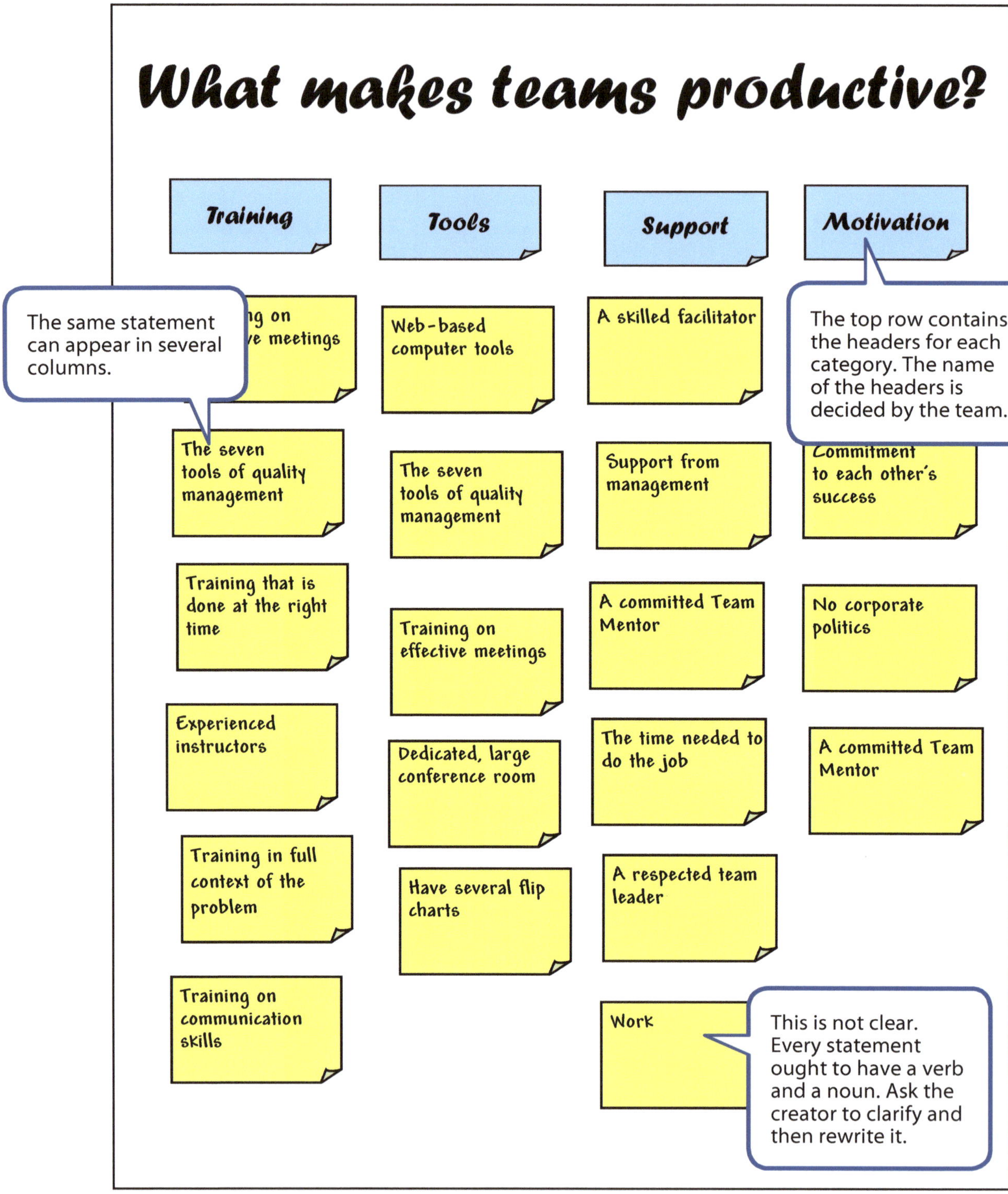

The Fishbone Diagram

Diagramming Cause and Effect

Generally, when something goes wrong — for example, a machine stops working — we tend to jump to conclusions based on our experiences and immediately infer the reason for the failure. Worse yet, we go all the way to a solution. There is a risk in this behavior. What if the reason for the breakdown is not what we thought? What if our solution is the wrong one?

The Fishbone Diagram, also known as the Ishikawa Diagram, is a fine tool for providing the discipline to go from "symptom" to actual "root cause" of the problem, regardless of the nature of the problem.

Teams are frequently confronted with solving problems; i.e., addressing a snag or an obstacle. Furthermore, the definition of the problem tends to be, in reality, a symptom of something else. For example, "the press does not work" or "the car does not start."

It is very tempting to try to solve the symptom of the problem without trying to analyze the real cause. In the example above, "the press does not work" may lead you to try several fixes, only to find out that the problem persists. Instead, taking time to characterize and prioritize the causes of the problem leads the team to find a permanent solution. This is where the Fishbone Diagram comes into play.

The cause-and-effect diagram is used to discover all the possible causes of a problem; those things that cause the effect; i.e., the symptoms. Further, it helps to prioritize the causes and drive a methodical identification of the true causes of a problem, thus leading to a permanent solution.

Why Cause and Effect

- You can't afford the trial and error to fix problems in your organization.
- It is a well-developed, mature process. It has been shown to be reliable and very practical.
- It taps the knowledge and experience of the entire team — a valuable source of knowledge and wisdom.
- It is user-friendly and requires no special math!

Definition

Kaoru Ishikawa is one of the pioneers of quality management processes. He was at the Kawasaki shipyards, where he became one of the founding fathers of modern management.

The Shape of Cause-and-Effect Diagrams

- A Fishbone Diagram is a methodical and graphical way to identify the root cause of a problem. It is a way to analyze symptoms, organize them and reach a logical conclusion about the cause. It is a way to plot cause and effect. The term "fishbone" is in reference to the appearance of these cause-and-effect diagrams because when completed they resemble the skeleton of fish, with the "head" being the problem statement, and the "bones" being the possible causes of the problem.
- The number of bones can vary from as few as four to as many as six, depending on the complexity of the issue or the major possible areas of the problem. Each major bone is a category of causes; for example, tools.
- Each major bone will generate sub-branches, driven by the thread of the conversation. For example, "the press does not work" may lead to causes such as "the on-off button is broken," which leads to "the contacts are burned out," which leads to "there was a power surge that burned out the contact." As you can see, there are many layers or levels depending of the depth of characterization.
- The number of layers or levels of each major bone depends on reaching the root cause. That is, when you have reached the point where you can diagnose the root cause of the problem, then you need no more layers.

The Fishbone Diagram

General Rules for Fishbone Diagrams

1. A useful Fishbone Diagram takes two to three hours to complete. Ensure that all participants have the time available. You should not rush it, as it relies on participation and brainstorming.
2. Encourage openness within your group. In some organizations, fear plays a critical role in stifling valuable input. You should reassure the participants that no one will be punished. This is called creating a safe environment. Your own behavior will have a significant impact in the degree of openness.
3. Begin by clearly stating the issue in the form of a problem. That is, you are starting with the effect and searching for the cause. For example, "The manufacturing line stopped at noon." Another example: "The phone rings too many times, and customers are upset."
4. "Why" is your best friend. Constantly asking, "Why?" will encourage the participants to better identify the causes.

The Steps

1. Always start with the effect in the form of a problem statement. For example, "There are too many rejects on station seven" or "The number of customer complaints has risen by 50 percent."
2. Encourage brainstorming to identify the cause of the problem. Reassure people that no one will be punished; it is safe to make suggestions, even if they are not 100 person certain. Encourage off-the-wall statements, and even inject humor and laughter. You must create a safe environment to enable all the information to flow freely.
3. Start by writing the input from the team members on a flip chart or marking board. Once you have enough ideas, start building the fishbone by categorizing the causes into clusters. Generally, there are four primary areas or clusters, although you may choose to have more. We discourage using more than six, as it becomes too difficult to manage. The four most common areas are equipment, people, processes and methods. For manufacturing situations, you might choose people, methods, materials and equipment. For service or business processes, you might choose equipment, policies, processes and people. Or you may replace those terms with ones that are more attuned with your culture and the nature of the problem.
4. Once you finish the categories and have exhausted all ideas, start prioritizing using your own criteria.
5. For complicated issues, you might want to select the top three or four, and have the team conduct additional research. This is particularly applicable in manufacturing or highly technical areas where validation is required.

Important

The criteria for prioritizing can be developed on the spot. For example, you might decide to use "1, 2, 3" to represent "most likely source of the problem," where 1 is the most likely. The criteria must be short and easy to understand.

Ask the team to help you prioritize using the newly developed criteria Assign responsibilities once all the prioritization is finished.

General Recommendations for Fishbone Diagrams

- Select the proper team — those people who have visibility of the problem.
- The phrasing of the problem statement is critical, as it sets the tone for the brainstorming. Carefully wordsmith the problem such that it clearly states the real issue. For example, "The press does not work" may not be sufficient. A better way might be to add more detail: "The press in station seven does not drill properly." We suggest that you work with your team to craft a statement that truly focuses on the real issue.

The Fishbone Diagram

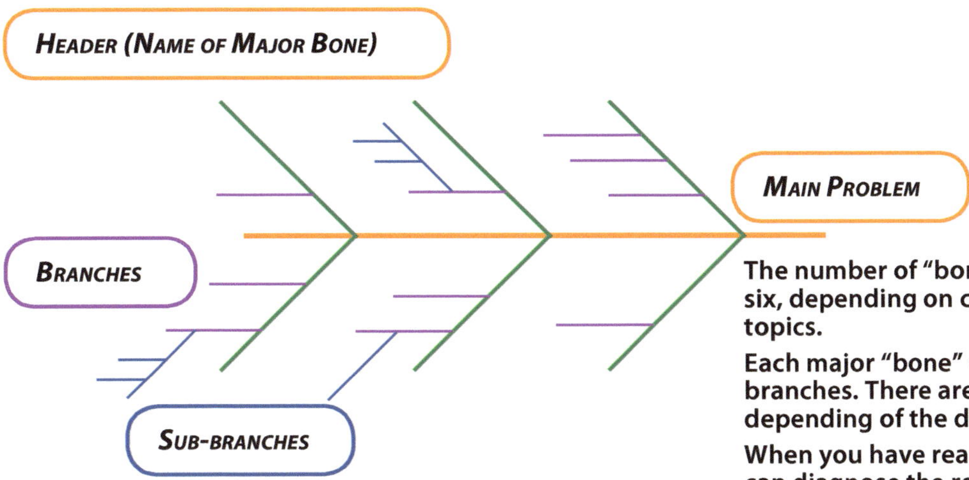

Cause-and-Effect diagrams resemble the skeleton of fish, with the "head" being the problem statement, and the "bones" being the possible causes of the problem.

The number of "bones" can vary from four to six, depending on complexity or number of topics.

Each major "bone" will generate sub-branches. There are many layers or levels depending of the depth of characterization.

When you have reached the point where you can diagnose the root cause of the problem, then you need no more layers.

Detailed Process: Fishbone Cause and Effect

1 Invite Participants

Who should attend?

- All stakeholders in the process or project. Try to keep the group to 15 participants or fewer.
- The ideal number is about 10 participants.

2 Prepare

Checklist:

- A large conference room.
- Post-It Notes® (3M brand, model #655 3x5).
- Sharpie® (Sanford brand, fine point, black markers) for all attendees.
- Flip chart paper and two easels.
- Masking tape.

3 Set up the Topic

- State the topic in the form of a problem or an issue. For example: "The technical support process does not work." Another example: "The scope of the new software application is not complete."
- Ensure that everyone understands the problem statement. You may need to rephrase it to ensure full understanding.
- Reassure everyone that there are no wrong answers.
- Reassure everyone that grammar and spelling are not important.
- Explain how to use the yellow stickies: one per subject, use the felt-tip pen, landscape orientation and NO criticism.

4 Start the Process

- Given the question or topic, ask everyone to write at least three yellow stickies outlining the reason the problem is occurring. For example, "Why did the technical support process break?" "It broke because there is no automatic answering software."
- When everyone is ready, ask each person to loudly and clearly read their yellow stickies.
- Tell everyone to listen intently as others read their yellow stickies.
- Encourage the participants to write additional yellow stickies as others present theirs.
- Collect all yellow stickies and randomly place them on the flip chart or board.
- Continue until the group runs out of ideas or repetition sets in.

Continued

The Fishbone Diagram

Cont'd

5 Start the Fishbone

- Give instructions to the group.
- Rally the group around the flip charts with the yellow stickies. The group may require encouragement to move up to the board.
- Draw the fishbone; that is, a line with four or six "ribs" emanating from it.
- Ask the participants to help you sort or cluster the yellow stickies into several topics. These topics represent each of the ribs.

6 Create the Categories (Rib Labels)

- After you have completed the sorting, label each of the ribs with a header. The header is a name that best represents why the yellow stickies in that particular rib are together.
- This is also a good time to scrub your information. For example, you can delete duplicate yellow stickies and you can rephrase or rewrite some yellow stickies to make them easier to understand.
- Important: some yellow stickies, will actually be a subset or very complementary to others. This represents a branch. So you will have the main ribs, and within each rib you may have branches. You guidance is to layer down to the root cause of the problem.

7 Prioritize

- It is best to prioritize the ribs (headers) rather than each sticky.
- Save the flip charts for archival purposes or to transcribe if it is important to do so.
- Debrief the group on the effectiveness of the tool, and ways to improve the affinity process.

See Also
Developing Scales for Prioritization

8 Develop Action Plan

- Assign responsibility for additional work; for example, more research may be needed to validate your findings.
- Develop a plan to ensure that your findings translate to action that will resolve the problem.
- It may be necessary to have additional meetings to continue the process.
- You may need to form sub-teams to address specific areas.
- Above all, leave no loose ends. Ensure closure!

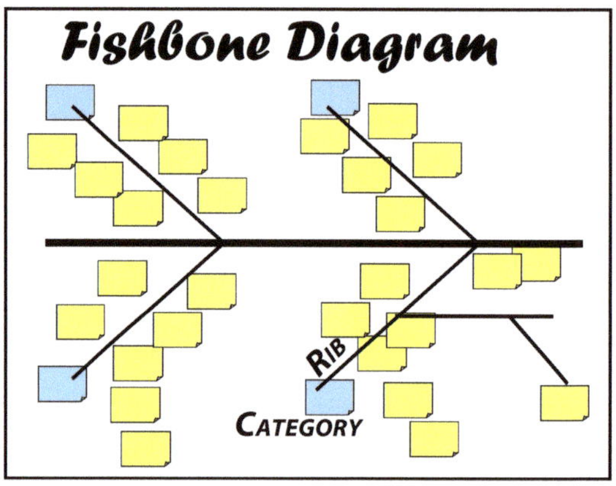

There should be from four to six ribs — don't go over six or it becomes too difficult to mange.

Example categories (rib labels):

- **People**
- **Processes**
- **Methods**
- **Tools**
- **Materials**
- **Equipment**
- **Information**
- **Policies**

The Fishbone Diagram

Important The Fishbone Diagram is only a tool to diagnose the problem, specifically the root cause of a problem. It does not solve the problem. For that, you need to assign corrective action. The diagram provides the framework to methodically diagnose a problem and then diagnose the most likely cause.

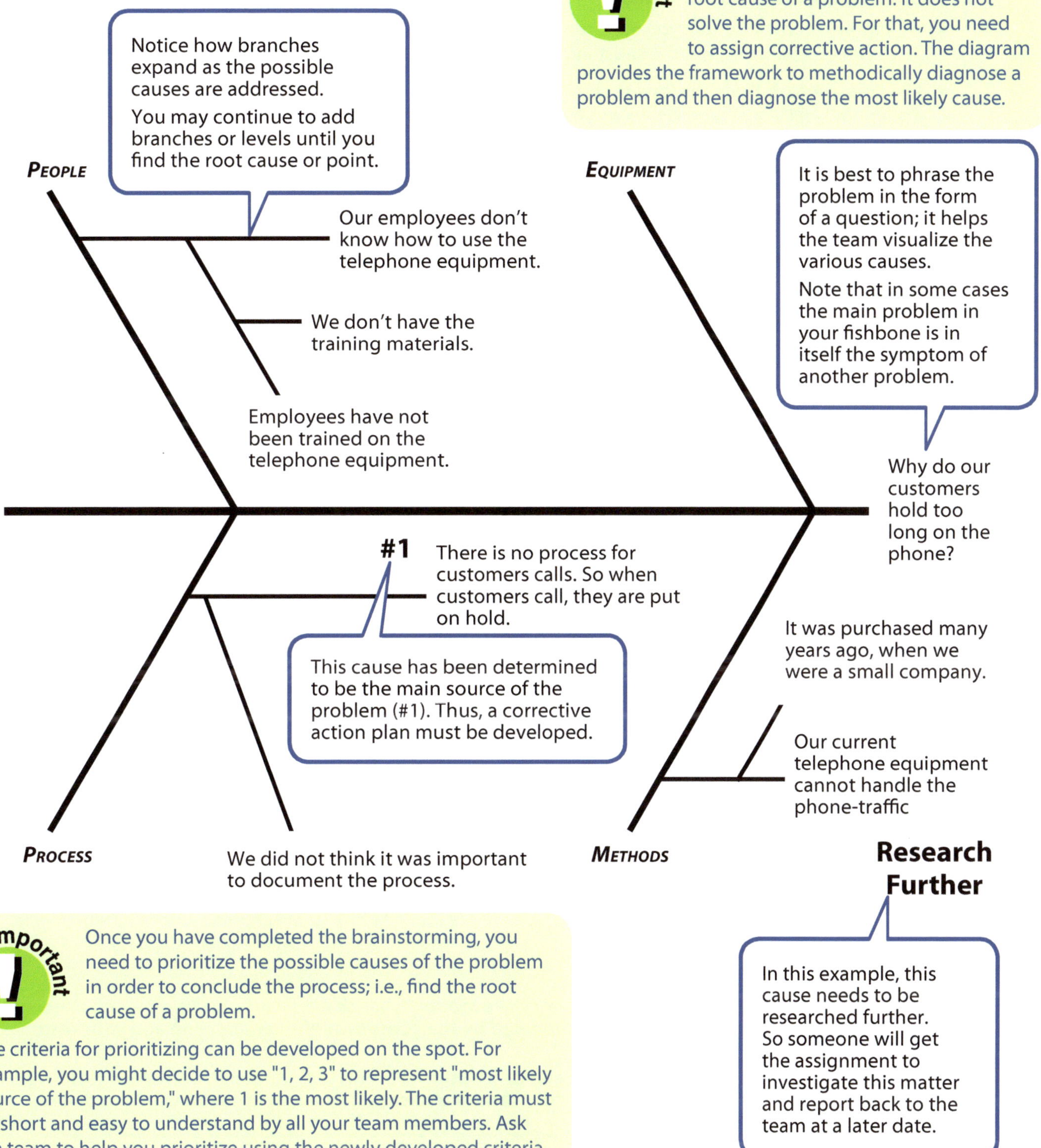

Important Once you have completed the brainstorming, you need to prioritize the possible causes of the problem in order to conclude the process; i.e., find the root cause of a problem.

The criteria for prioritizing can be developed on the spot. For example, you might decide to use "1, 2, 3" to represent "most likely source of the problem," where 1 is the most likely. The criteria must be short and easy to understand by all your team members. Ask the team to help you prioritize using the newly developed criteria. Assign responsibilities once all the prioritization is finished; for example, take all the "1's" and assign responsibilities.

The Fishbone Diagram

Developing Scales for Prioritization

Recommendations and Rules

- We highly recommend that you use simple, linear scales from 1 to 10 in every instance. This will keep the focus not on the scales, but the substance of your risk management efforts.
- Always use 1 to indicate the lowest or minimal level, while 10 will always indicate the highest or maximum level. For example, maximum damage or impact of a risk is always 10. Correspondingly, 1 will always indicate the lowest impact. It follows that the probability of a risk occurring will also be highest at a 10 and lowest at a 1. This approach will allow you to obtain information in a harmonized, or aligned, way.
- Refrain from using scales that are too complicated, such as formulas, weighted averages or algorithms. Keep in mind that this is a **QUALITATIVE** effort and adding more resolution will not add more accuracy; on the contrary, it may confuse your team.
- You and your team need to define each step in your scale — for example, 10 means Highly Likely, 8 means Likely, 6 means Moderate, 4 means Occasionally, 2 means Remote and 1 means Unlikely. You must do this in order to allow everyone in your team to have the same perception of a level. Experience will help you decide the amount of definition detail that each level requires.
- We suggest that you start by developing the description of the levels using the even numbers in your scale. This will save you time. As you become more experienced, you can complete the definitions from 1 to 10
- Be certain to document your scales and the definition of the levels. This will allow for repeatability, reference and continuous improvement.

Example: Simple Scales and Definition Levels

A hypothetical scale for probability might describe the levels as shown in the figure below:

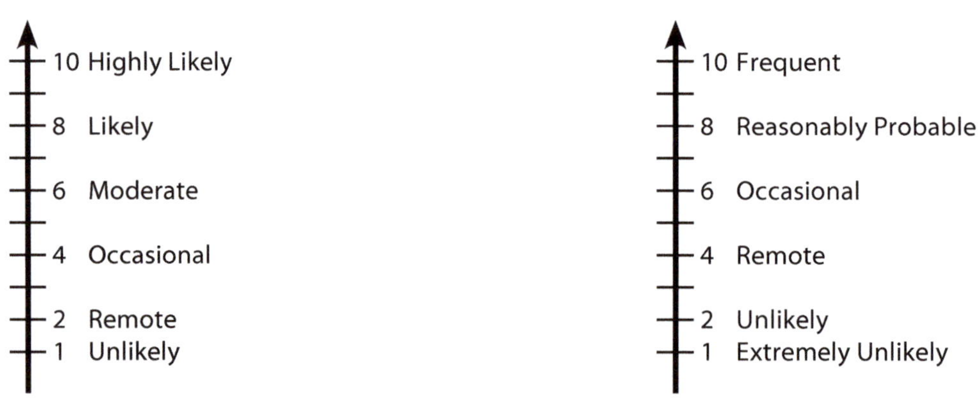

Here are two very simple definitions of a scale for the probability a risk might occur. The meaning of each level (step) of the scales would have been determined by the consensus of the team. You and your team could use either scale as a starting point.

Over time, you can add additional definitions and descriptions among the different levels. Your team will become familiar with the meaning of each number, and the uncertainty between levels will be reduced.

Example: Sophisticated Scale and Definition Levels

A hypothetical scale for impact or damage might describe the levels as shown in the table below: Over time, you can develop scales that are much more meaningful to your situation.

Level	Rank	Severity of Failure as Seen by the Customer
1	None	No noticeable defects.
2	Very minor	Abnormal noises, noticeable only by the most discerning operators.
3	Minor	Abnormal noises, noticeable by 50% of the customers. Fully operational, but causes stress on the operator.
4	Very low	Abnormal noises, noticeable by the customer. Fully operational, but causes stress on the operator.
5	Low	Operable but difficult to operate, all functions available.
6	Moderate	Operable but difficult to operate, some functions not available.
7	High	Operable, but at reduced level of performance.
8	Very high	Device inoperable.
9	Hazardous with warning	Unsafe to operate.
10	Hazardous without warning	Operator injury very likely.

Here is an example of a more sophisticated definition of a scale, one that focuses on the levels of impact for each risk. The meaning of each level (step) of the scales would have been determined by the consensus of the team.

Appendixes

Appendix A: Recommended Books 145

These lists reference books that the senior consultants at Rapidinnovation, LLC have read and found applicable

Appendix B: Glossary 152

Terms you will encounter in the real world of product development explained in a practical, common-sense fashion. With this resource, you will be able to better connect the concepts and tools of product development to your projects.

Appendix C: FMEA Process Map from SEMATECH, Inc. 175

Appendix D: Templates 176

These templates are for you to use and share. Throughout this guidebook, you have seen real world tools and examples. Any page with a template icon indicates that there is a correlating fillable form in this section.

Recommended Books

Introduction

These lists reference books that the senior consultants at Rapidinnovation, LLC have read and found applicable to today's business challenges. Rapidinnovation, LLC is a consulting firm whose primary focus is serving product development organizations around the world, which we've been doing since 1997, and with many satisfied clients.

We welcome your recommendations and reviews of these and other books and resources. Please e-mail the authors with your comments and suggestions.

Teams and Teamwork for Product Development

Book Title & Author	Subject	Comments
The Wisdom of Teams By Jon R. Katzenback Published by Harvard Business School Press	• High-performance teams.	• Excellent book for creating development teams. • Clear, no-nonsense approach to cross-functional teams such as concurrent engineering, product development, etc.
Peopleware, Productive Projects and Teams, Second Ed. By Tom DeMarco & Timothy Lister Published by Dorset House Publishing	• Software development. • People management.	• Good book to help understand the human side of software development.
Principle-Centered Leadership By Stephen R. Covey Published by Fireside	• Timeless principles of leadership. • Clear, no-nonsense understanding of leadership in a business environment.	• The difference between a manager and a leader, and its implications in an organization.
Punished by Rewards By Alfie Kohn Published by Houghton Mifflin Company	• Reward, recognition, celebrations and motivation in organizations.	• The truth about failed recognition programs and the unintended results of poor motivation and recognition programs.
Mastering Virtual Teams, Third Ed. By Debora L. Duarte and Nancy Tennant Published by Jossey-Bass	• The ways and means for enabling virtual teams to succeed.	• Duarte does a good job of presenting the challenges and solutions for distributed teams. A level-headed presentation of the issue.

Recommended Books

General Management and Strategy

Book Title & Author	Subject	Comments
The Innovator's Dilemma By Clayton M. Christensen Published by HarperBusiness	• Why companies fail to stay at the top of their industries.	• A book that confronts a reality faced by most successful technology companies: the need to change, and change fast.
The Discipline of Market Leaders By Michael Treacy Published by Perseus Books	• Choose your customers, narrow your focus, dominate your market.	• Deliberately choosing the customers you wish to serve allows you to develop the competencies needed to succeed.
Built to Last By James C. Collins Published by HarperBusiness	• Visionary companies. • Values, culture and successful businesses.	• Focusses on the human side of corporate culture for the promotion of long-term success.
The Balanced Score Card By Robert S. Kaplan and David P. Norton Published by Harvard Business Press	• What to measure and why. • Financial measures alone are not enough to ensure sustainable customer satisfaction and profits.	• Rational approach to measures. • Systemic look at the enterprise. • Focus on the human side of measures.
Companies with a Conscience, Third Ed. By Mary Scott and Howard Rothman Published by Citadel Press	• Case studies of 12 companies that are profitable and at the same time people-, environment- and community-friendly	• Good examples of "having your cake and it eating it too" — the ability to make a profit and have a clear corporate conscience.

Product Development

Book Title & Author	Subject	Comments
Rapid Development: Taming Wild Software Schedules By Steve McConnell Published by Microsoft Press	• Perspectives and methods for rapid software development. • Focus on the process and framework for development.	• The methodical approach to rapid software development. • Good food for thought.
Revolutionizing Product Development By Steven C. Wheelwright Published by The Free Press	• One of the classics on product development.	• A must-have for those who are serious about product development.
The Invisible Computer By Donald A. Norman Published by The MIT Press	• Why good products can fail.	• Insightful as to why technology companies behave like technology companies.
Dynamics of Software Development By Jim McCarthy Published by Microsoft Press	• Rapid software development is not the result of countless hours of coding.	• A book more about teams and teamwork than software development.

Recommended Books

Book Title & Author	Subject	Comments
Portfolio Management for New Products, Second Ed. By Robert G. Cooper, Scott J. Edgett and Elko J. Kleinschmidt Published by Perseus Books	• Managing of technology for profits. • Disciplined product development. • Separate the losers from the winners.	• A solid, direct book on managing a portfolio of products while optimizing resources. • Good methods for balancing your product offering.

Productivity Improvement

Book Title & Author	Subject	Comments
The Fifth Discipline By Peter M. Senge Published by Doubleday	• System thinking. • The quintessential book of systemic approaches to solve business problems.	• A must-read to understand the total solution to customers' problems • Also very important to re-engineering
Improving Performance: How to Manage the White Space on the Organization Chart By Geary A. Rummler Published by Jossey-Bass	• Business processes. • Teamwork in organizations. • Process mapping for non-manufacturing organizations.	• Clear understanding of processes in a business environment. • Good how-to information and a level-headed approach to organizational development.
Kaizen: The Key to Japan's Competitive Success By Masaaki Imai Published by McGraw Hill	• Continuous improvement.	• The truth about continuous improvement as a culture and not a fad.

Recommended Books

Project and Program Management

Book Title & Author	Subject	Comments
Flexible Project Management for Product Development By Jose Campos Published by Rapidinnovation, LLC	• Clear understanding of the conventional project management rules and tools do not necessarily work in a product development environment. But here, the author has experience in both project management and product development.	• Easy to understand and very detailed. • Plenty of examples, templates and forms ready to use. • Easy to share with other members of the development team. • Clearly written with product development in mind, very practical.
Project Management Toolbox By Dragan Milosevic Published by John Wiley and Sons	• A comprehensive set of project management tools full explanations and plenty of examples	• A must-have reference
Program Management for Improved Business Results By Dragan Milosevic, Russ Martinelli and James Waddell Published by John Wiley and Sons	• A business approach to the role of program managers. Authored by Dragan Milosevic, acclaimed authority in project and program management.	• Probably the best book on program management.
Reinventing Project Management By: Aaron J. Shenhar Published by Harvard Business Press	• Proof positive that every project is different and therefore, you cannot manage them exactly the same.	
The Fast Forward MBA in Project Management, 2008 Ed. By Eric Verzuh Published by John Wiley and Sons	• Looks at the challenges of today's project environment and examines the skills a project leader needs to make a project successful. • It includes organizational changes, tools and techniques, control methods and real-life applications.	• Good medium-depth overview of all aspects of project management. • Easy to read, with examples and case studies, but meaty enough that both junior and experienced project leaders will find nuggets. • Down-to-earth
The New Project Management, Second Ed. By J. Davidson Frame Published by Jossey-Bass	• Frame makes a case that conventional wisdom is only marginally relevant to project leaders working in today's turbulent project environments. • He offers strategies and techniques for changing from implementor to initiator, and for dealing with the soft and complex issues that affect most technology projects.	• Good ideas, especially on soft skills and orientation toward the customer. • Frame sometimes undervalues the traditional techniques and gets too enthusiastic about how empowered a project leader can be. Pair this book with a solid grounding in the hard fundamentals of project management, such as from Verzuh's book.

Recommended Books

Book Title & Author	Subject	Comments
Visualizing Project Management, Third Ed. By Forssberg, Mooz and Cotterman Published by John Wiley and Sons	• A recently updated project management textbook with lots of examples. It strives to provide an integrated model of project management, and focuses on vocabulary, teamwork, the project cycle and management techniques.	• Covers a wide variety of hard and soft project management topics, with short case studies of how they have been used in real situations. • Focuses mostly on applying them within a sequential framework, so don't look for lots of insight on iterative approaches here. The authors' experience in aerospace and government projects shows through in the vocabulary and examples.
The Wiley Guide to Managing Projects By Peter W. G. Morris and Jeffrey K. Pinto Published by John Wiley and Sons	• Published in 2005, this book is up-dated every 10 years. This is a comprehensive book on project management. Looks well beyond the basics on project management (cost, schedule and performance).	• A must-have reference for those who are serious about project management. • Critically important for those involved in the strategies of project management.

The Voice of the Customer (VOC)

Book Title & Author	Subject	Comments
The Voice of the Customer for Product Development By Jose Campos, Jean-Claude Balland Ph.D. Published by Rapidinnovation, LLC	• A detailed guidebook on how to capture and use the voice of the customers (VOC) for product development. The author describes every step and detail in an easy to understand fashion.	• Easy to understand and very detailed. • Plenty of examples, templates and forms ready to use. • Easy to share with other members of the development team.
How Customers Think By Gerald Zaltman Published by Harvard Business School	• Essential insight into the mind of the customers and the market.	• Customer needs are much more subjective than we thought. • An eye opener concerning value.
Advanced QFD By Larry Shillito Published by Wiley & Sons, Inc.	• The use of QFD (Quality Function Deployment) in a real-world environment, linking technology to the market.	• A sound, levelheaded book on QFD, A must-have if you are serious about QFD.
Software Requirements, Second Ed. and More About Software Requirements By Karl Wiegers, Published by Microsoft Press	• A practical guide for gathering and managing requirements throughout the product development cycle.	• As relevant to product requirements as it is to software requirements, this practical guide will go a long way towards making your R&D and Marketing people see the world through the same lens.

Recommended Books

Book Title & Author	Subject	Comments
Customer-centric Product Definition By Sheila Mello Published by AMACOM	• Good coverage of the Voice of the Customer (VOC) and its use in defining new products.	• Very good book; the author clearly portrays the process of getting the customers at the center of new product development.
Writing Effective Use Cases By Alistair Cockburn Published by Addison Wesley	• A practical guide to the powerful approach of expressing the behavioral requirements of complex systems.	• Practical, with numerous examples of good and bad use cases.
Customer Visits, Second Ed. By Edward F. McQuarrie Published by Sage Publications	• A clear description of what it takes to perform a good customer visit.	• The author is well-recognized as an authority in this field. Levelheaded approach.
Acquiring, Processing, and Deploying the Voice of the Customer By M. Larry Shillito Published by St. Lucie Press	• Description of ways to obtain, process and use the voice of the customer.	• Plain English, down-to-earth, but clearly from someone who knows.
Voices Into Choices By Gary Burchill Published by Joiner	• A handbook with step-by-step descriptions to obtain and use the voice of the customer	• Comprehensive. • Truly a fine work — a must-have handbook.
Priceless: Turning Ordinary Products Into Extraordinary Experience By D. LaSalle & T. A. Britton Published by Harvard Business Press, 2003	• The title says it all. People don't buy products, they buy value and experience. This book will help you see the world through the eyes and heart of customers.	• Very valuable in bringing the new concept of experience to a tactical level.
Quality Function Deployment: How to Make QFD Work for You By Lou Cohen Published by Addison-Wesley, 1995	• A tour of QFD fundamentals, as well as advanced concepts	• Clear exposé to a sophisticated subject. • Most useful to engineers than marketers.
The Observational Research Handbook: Understanding How Consumers Live with Your Product By Bill Abrams Published by McGraw-Hill Professional, 2000	• Practical coverage of the art of observing customers in context to understanding latent needs.	• Useful advice on how and when to observe customers to get clues to their needs.
Learning from Strangers: The Art and Method of Qualitative Interview Studies By Robert S. Weiss Published by Simon & Schuster, 1995	• The subtitle of the book says it all: "The art and method of qualitative interview studies." These are important skills to master.	• A how-to book on interviewing customers by an expert on the topic.

Recommended Books

Risk Management and FMEA (FMCA)

Book Title & Author	Subject	Comments
Risk Management and FMEA for Product Development By: Jose Campos Published by Rapidinnovation. LLC	• A detailed guidebook on risk management for product development. The author describes every step and detail in an easy to understand fashion.	• Easy to understand and very detailed. • Plenty of examples, templates and forms ready to use.
Waltzing with Bears: Managing Risk on Software Projects By Tom DeMarco and Timothy Lister Published by Dorset House Publishing		
Proactive Risk Management: Controlling Uncertainty in Product Development By Preston G. Smith and Guy M. Merritt Published by Productivity Press	• A "plain English" book on risk management.	• A good reference for anyone responsible for managing risk.
Managing the Unknown By Christoph H. Lock, Arnound DeMeyer and Michael T. Pich Published by John Wiley & Son	• A book about managing the risks you don't know: the ways and means to deal with total uncertainty. • Intriguing new concepts about risk, particularly in areas with high uncertainty such as new technology.	• A good book to have if you are in product development or technology development. • Challenges conventional views of risk management.
Project and Program Risk Management: A Guide to Managing Project Risks and Opportunities By R. Max Widerman Published by Project Management Institute, 1992	• A risk management book for beginners and intermediate managers of risk.	
Failure Mode Effect Analysis: FMEA from Theory to Execution, Second Ed. By D. H. Stamatis Published by American Society for Quality, 2003		• A comprehensive book on FMEA, strong in context and advanced methods. • A good overall reference book for those serious about FMEA.
Project Management Tool Box By Dragan Milosevic Published by John Wiley and Sons, 2003	• A comprehensive set of project management tools, full explanations and plenty of examples.	• A must-have reference.

Glossary

APPENDIX B

Definitions for Product Development

Practical Definitions in Plain English to Enable Effective Action

The intent of this glossary is to explain the terms you will encounter in the real world of product development. Furthermore, we explain these terms in a practical, common-sense fashion. When multiple relevant definitions exist, we will provide them. With this resource, you will be able to better connect the concepts and tools of product development to your projects.

Term	Definition & Notes
Agile	• A software development model or framework based on frequent iterations with the customer and development of the software in small, incremental amounts, collaboration and flexibility.
Alpha Unit	• A version of a new product that is ready for internal testing. The product meets criteria that enable testing that will yield valuable data and input to proceed with the final version. • After the Alpha Unit comes the Beta Unit, a more mature version of the product. • In hardware an alpha unit may be used to test the manufacturing process, functionality or testing strategy. • In software it can be used to test human interface (GUI), usability or functionality. • Alpha units are not saleable and are intended for internal use only.
Assumption(s)	• A short set of relevant factors that can have a positive or negative impact on your project or program and define the playing field. – The playing field refers to the arena or area of focus for your endeavor. – A short set means three to seven relevant factors. – Relevant means of importance to you and your team — that which must be taken into consideration when making decisions.
Attribute	• Used to define a set of general characteristics of a platform, product line or product, an aggregate of several features. Examples include "ease of use," "ruggedness" and "re-configurable". • Related to, but not the same as "Product feature." See also "Feature."
Awareness	• A measure of the percent of target customers who are aware of the existence of something. It could be a product, a brand or even a concept. • Awareness does not imply preference. • Awareness can be measured for brand, product or any other attribute you select. • Awareness should be measured before and after an event; for example, an advertising campaign. The difference provides a measure of effectiveness. • There are two kinds of awareness: Unaided awareness is when the customer volunteers the information (manifests the awareness) without any help. Aided awareness is when a hint or suggestion is provided, such as a photograph, and then the customer expresses awareness.

Glossary

Term	Definition & Notes
Benchmarking	• The process of studying how successful companies, industries and even competitors perform specific functions and then selecting the best as standards of performance for your own improvements. • You can benchmark with your competitors or across industries, in your own country or across the world. • Benchmarking is a good way to develop your level of excellence, and to learn "good practices" for your continuous improvement.
Best Practices	• A method, technique, tool or other practice that has shown to be the best within a certain category. Best practices must be researched, defensible and published by an authorized organization. • Methods, approaches or processes deemed to be "best in class"and, consequently, worthy of imitation. • See also "Good Practices."
Beta Test (a.k.a. Beta Unit or Beta Version)	• An external test of pre-production products. A way to test new products in actual application to discover bugs and last minute fixes before full introduction. • Some companies (software in particular) use Beta Tests as part of their introduction strategy, to generate awareness and word of mouth.
Bill of Material (BOM)	• A hierarchical list of systems, sub systems, sub assemblies and components that make up a product or a higher level sub-system. • It can be a set of parts list, drawings, and schematics. In manufacturing it can also mean the way a new product is assembled including all parts, systems and sub systems.
Brainstorming	• The process of tapping into a group's creativity to generate ideas, solutions or innovation. • Generally, it takes the form of a meeting or gathering where the team or group is encouraged to come up with as many creative ideas as possible. • There are many methods to facilitate or enable brainstorming. The methods are intended to enable the ideation, processing of the information, and finally prioritization of the ideas. Many brainstorming sessions fail due to a lack of a framework or process.
Brand	• The combination of all tangibles, intangibles and actions of a company that create a particular promise of value in the customers' mind. • A brand is a holistic concept that evokes the mental images and expectations the customers have about a particular company.
Breakeven	• A financial metric that identifies the point where the revenue (sales) of a new product match the investment made in developing it. • At this point, the new product has reached enough sales to start generating "positive cash flow."

Glossary

Term	Definition & Notes
Business Case	• The results of the business analysis to demonstrate the viability of a new product or service prior to approval for development. • Viability includes profitability, market share or other critical and strategic objectives of your company. • "Making the business case" in product development implies your ability to convince the company executives that investing in your new product will result in a measurable return. Philosophically, you are going to the company executives to ask for an investment, which will results is a quantifiable return
Business Model	• The ways and means selected by the company to generate revenue and make a profit. • The Business Model is said to be working when the company has deployed and executed the strategies to generate revenue, profits and serve its customers.
Business to Business (B-to-B)	• Sales to other businesses and not to individual consumers. For example, an OEM sells components to a manufacturing company. • Sales to distributors, representatives or wholesalers are also considered B-to-B.
Champion	• A person who takes strong personal interest in seeing that a particular process, product or other is carried to fruition. Champions are always ready to deal with adversity and committed to success. • Champions are said to be "maniacal" in their commitment, willing to put exceptional effort and energy into their cause. • No innovation can be brought to fruition without champions.
Change Management	• An organizational process aimed at helping employees embrace changes in their current business environment and then change their behavior in support of the new state. • A structured approach to shifting or transitioning individuals, teams and organizations from a current state of behavior to a desired future state.
Charter, Team Charter	• The document that officially starts an initiative. This document outlines the team members, the common purpose that brings them together and the performance goals to ensure progress and outcomes. It also defines the roles, responsibilities and authority. It defines the desired approach to ensure completion. • The team members should be strongly involved in — even responsible for — the development of the charter. • The Charter must be fully supported by Management. • A charter is the unifying principle of the team. Without it, each team member has a different idea of what needs to get done and how.
Co-location	• Product development where the entire development team is in one location; for example, the same building. • The processes and methods for co-location development are different from that used for multi-site or distributed product development.
Commercialization	• The process of taking a new product to market. The transformation of an idea into a product that customers are willing to purchase. • Commercialization can be used for a new product, a new technology or anything that a particular set of customers value and are willing to pay to obtain it.

Glossary

Term	Definition & Notes
Compelling Reason To Act (CRTA)	• The needs the customer must address now so that a budget and the will to act can be generated. • CRTA is proposed to designate the highest priority needs — the ones that can be funded and will be acted upon by your customer. • A customer must have a CRTA in order to purchase your product. It is your responsibility to identify CRTAs so you can develop products that address customer requirements.
Competency	• The aggregate set of skills, knowledge and abilities that an individual needs to fulfill a specific set of measurable responsibilities. • Competency is all about one individual — not a team or an organization. The equivalent statement at the organizational or team level is known as "maturity" or "organizational maturity"
Complete Product	• The aggregate of the tangible, intangible and emotional components that constitute a product in the eyes of the customer. • See also "Offering" and "Whole Product."
Completion	• The verifiable fulfillment of all objectives. • The conclusion of all efforts; for example, the completion of a new product development is the introduction or rollout to the market place.
Compound Annual Growth Rate (CARG)	• The calculation of rate at which an investment would grow if it grew at a steady rate. It's a way to smooth the return on an investment for the purposes of making informed decisions. • In product development, CARG is generally used to describe the growth rate of a market or the growth rate of new products.
Computer Aided Design (CAD)	• The use of computers and software to design a new product or parts of a new product. • CAD is a very advanced technology enabling simulation, testing, and all aspects of design.
Computer Aided Manufacturing (CAM)	• The use of computers and software to design and run a manufacturing process in an integrated fashion; that is, as a system. • In some areas CAM is also used to describe any aspect of the manufacturing process that is run with the aid of computers or fully automatic, not necessarily the entire system of manufacturing. • Widely used in industry, CAM has become essential to running manufacturing operations around the world. • CIM (Computer Integrated Manufacturing) may be used in conjunction with CAM.
Computer Integrated Manufacturing (CIM)	• See "Computer Aided Manufacturing (CAM)"
Concurrency	• The ability to develop several products at the same time. • The goal of a product company is to achieve speed, concurrency and repeatability. • It can also be applied to the development of a single product where parts, assemblies and subassemblies are developed at the same time and not sequentially.

Glossary

Term	Definition & Notes
Concurrent Engineering	• A model of product development where all functions such as engineering, marketing, manufacturing, etc., work at the same time to design and take the product to market. • A set of methodologies that enable rapid product development by allowing as many activities as possible to take place simultaneously. • Contrasted to the sequential model, where one activity had to wait for another before it could be started.
Concurrent Engineering	• An emerging engineering methodology based on the premises: – All aspects of the life-cycle of a product should be considered as part of the initial design process; that is, from basic functionality until end of life. – All activities that can happen at the same time must be undertaken. • The methodology, if properly applied, yields better products in a shorter amount of time.
Conjoint Analysis	• A quantitative market research technique used to characterize how people value the features (or benefits) of a product. • One of the few methods available to determine the possible price of a new product with objectivity. • Conjoint analysis is powerful because it puts the customer in a real life situation of having to make trade-offs between benefits (what he wants to get) and costs (what he has to give to get it).
Constraint	• Restriction or boundary imposed on a project or design. • Constraints may help in setting priorities, avoiding risks or the physical limitation of the technology.
Contextual Inquiry	• A way to gain knowledge about a customer by observing people in their environment doing specific activities. • The use of anthropology, social science and other techniques to study customers performing specific tasks for the purpose of understanding their needs. • The observation must be qualitative.
Core Competency	• The few things that a company does that: – Its customers find exceptionally valuable. – It does better than its competitors. – Its competitors find difficult to imitate. – It would never outsource to anyone. – It can transfer to other industries as is. – Is part of its growth plan. • The definition of Core Competency varies. It's driven by the specific situation of a company at the time it defines its Core Competencies. Nonetheless, it is critical for a company to understand and develop its core competencies. • Core competencies are always a web of things that a company does particularly well without sometimes even be conscious of it.

Glossary

Term	Definition & Notes
Core Team: Concurrent Engineering	• A small group of representatives from the critical functions (e.g., Engineering, Marketing, Program Management, Finance, etc.) responsible for the development of a specific product and accountable to deliver measurable results. • The Core Team is responsible and accountable for progress and eventual delivery of the new product to the market. It acts as the Program Manger for the development.
Core Team: Generic	• A small group of three to five individuals that is chartered and empowered to guide the implementation and completion of an initiative. • Very useful in fast-changing environments where distractions abound because a small group can focus attention and energy to ensure clean execution.
Corrective Action	• The specific steps to eliminate the cause of a detected nonconformity or other undesirable situation. That it, the steps that are taken to permanently remove the causes of an existing nonconformity through root cause analysis and deploying permanent corrections. • In contrast, "Remedial Action"refers to the temporary steps taken to eliminate one specific non-conformance without regards to a permanent solution.
Criteria	• A standard used to evaluate alternatives in a decision. Criteria are always finite and easy to understand by all. • In the product development process, the criteria are generally defined by the Gatekeeper at major gates.
Critical Path (CP)	• The path with the longest duration • CP comes from the analysis of a network diagram or a PERT (Project Evaluation and Review Technique) diagram in project management.
Critical Success Factor (CSF)	• Something that is critical to your program and must happen. • The few key areas where things must go right for the business to flourish. If results in these areas are not adequate, the organization's efforts for the period will be less than desired. • Those few things that must happen in order to achieve the desired results.
Cross Section (Market)	• A typical selection or sample. For example, a cross section of a particular market segment. • This term is a metaphor from engineering, used to reflect the framework of a particular structure such as a market segment.
Cross-Functional Team (a.k.a Multi-Functional Team)	• A team made up of representatives from various functions. For example, marketing, engineering and manufacturing. • Most product development is achieved though cross-functional teams.

Glossary

Term	Definition & Notes
Culture	• A mosaic of interrelated elements and their collective day-to-day interaction that shape the behaviors of people in the group and what's right and wrong. • For example: – Management behavior. – Customs, myths and norms. – Ceremonies, rituals and events. – Rewards and consequences. – Physical environment. – Rules and policies. • A company culture develops when corporate values are consistently followed for a number of years. • Said another way: "Culture Consists Of A Shared, Commonly Held Body Of General Beliefs And Values That Define The 'Shoulds' And The 'Oughts' Of Life." — Kluckhohn
Customer	• The person or entity that pays for the value of a product or service. • There is tendency in some organizations to refer to teams and functional groups as "internal customers." This tends to diffuse the importance of the one and only true customer, the one that pays. • Customers can be "users" or not of your product. Make the distinction between the person who pays and the person who uses.
Customer Experience	• A succession of events in the life of the customer dealing with an offering. This interaction will generate good or bad feelings. • Designing experiences is the new frontier in user-centered design because creating pleasant feelings is the best way to generate satisfaction, loyalty and word of mouth. • Thinking in terms of customer experience forces the marketer to consider the entire sequence of events that the target customer undergoes when dealing with a company and its offerings.
Customer Intimacy	• The business definition as introduced by Michael Treacy in Discipline of Market Leaders is (paraphrase): The discipline of delivering what each specific customer wants or needs rather than what the market as a whole wants or needs. • Customer-intimate companies develop deep relationships with their customers so as to anticipate their needs and solve their problems better and faster.
Customer Requirements	• The crystal-clear articulation of the desired outcomes a customer is striving to achieve.
Customer Satisfaction	• The positive or negative evaluation or feeling of customers that results from comparing the actual experience with a product or service with the expectations it had prior to purchase and use. • Customer satisfaction is a different concept than "customer value." A customer may be satisfied or not with the value it gets.
Customer Value	• The perception of what customers will get (benefits) for what they have to give (costs). • Alternatively, it is the customers' perception of the net benefits they will derive from your offering at a particular price, situation and use.

Glossary

Term	Definition & Notes
Customer Value Drivers (CVDs)	• The criteria that are perceived by the customer to be the most important to the buying process. The yard stick that the customer uses to make the buy decision. • Customer value drivers can be tangible (objective) or intangible (subjective).
Customer-Driven	• The urge to grow a business culture committed to the relentless creation of superior value for the customer. • It is a set of values that put the customers' interest first in order to develop a long-term profitable enterprise — while not excluding those of other stakeholders (paraphrased from: Narver and Slater). • In a for-profit business it is part of the overall strategy to generate revenue and profits. A company can make a profit and not be customer-driven.
Defect	• Lack of conformance to a standard. • Since the introduction of Quality methods, a defect is anything that your customer says it is. This is an effort to focus the organization to evaluate products and services as the customer would.
Defects: Engineering	• Engineering "defects" include design errors, software bugs, incorrect documentation and any substandard design or development steps that would not occur if the process were under control.
Deliverables	• The visible or "touchable" results from a task or a project. The outcomes from a task or a project. • Deliverable has come to mean more than "able to be delivered", now more at "the results of an endeavor or a project." • For example, the deliverable from the marketing organizations is a Marketing Requirements Document (MRD).
Delphi Process	• A technique that uses iterative rounds of consensus development across a group of experts to arrive at the most probable outcome for some future state. • "Experts" are people who have knowledge in a particular area; for example, knowledge about a specific market segment.
Derivative Product	• A new product based on changes to an existing product without affecting the basic product architecture or platform. • A derivative product can be an improvement to an existing product or a different version to satisfy a specific set of customers. • In software development, a rev includes new value to the customer. A rev to fix bugs is not a derivative product.
Design Change Order (DCO)	• A process to ensure the smooth and controlled change of a design and minimizing disruption. • Documented changes to the design of a new process, which is part of a formal process for changes. • The purpose of the DCOs is to minimize disruptions to the design process and ensure clarity of communication. • Also used in this context: – ECN: Engineering Change Notice – ECC: Engineering Change Control – ECO: Engineering Change Order

Glossary

Term	Definition & Notes
Design Engineer	• A person with proven domain expertise and experience who is responsible for creating innovative solutions that customers value. • The output of a design engineer is a package of technology in the form of documentation (drawings) that is the tangible manifestation of an innovative solution that customers value.
Differentiated Product	• Characterizes a product that provides more advantages than its competitors for the most important Customer Value Drivers of a given segment. • A well-differentiated product is one of the most important success factor for a new product.
DOA (Dead on Arrival)	• Products that arrive at a customer site unusable. • A product that does not work when the customer unpacks it.
Early Adopters	• Customers who buy into a new product or technology very early in its life cycle. • Early Adopters are willing to take the risk of an unproven new product or technology to gain a business or personal advantage.
Economic Value Analyses (EVA)	• A methodology that enables you to understand the real progress of a development process by measuring actual progress towards completion. A way to clearly understand progress towards the objectives of a new product. • EVA is found in project management texts — very popular in the construction industry — and now being adopted in product development due to its objectivity. • While budge and schedule are indicators of "movement", they do not necessarily show progress towards the achievement of the objectives of the new product.
Elevator Pitch (a.k.a. 30-Sec. Commercial)	• A short statement summarizing in a compelling way a value proposition. • If a pitch can't interest a prospect in the time it takes to move a few floors in an elevator, you don't have a strong, well-differentiated offering. Rework the pitch and try again.
Engineering Change Control (ECC), **Engineering Change Notice (ECN),** **Engineering Change Order (ECO)**	• A formal process that documents a needed design change regardless of the root cause of such change. The process must follow sound engineering principles for control, tracking and disposition. • See "Design Change Order (DOC)." • EISs may include modeling, simulation, analysis and other computational requirements.
Engineering Information System (EIS)	• Applications and solutions for information-intensive engineering and scientific endeavors. The ability to model, process and manage very large amounts of information for the purpose of enabling engineers and scientists to accomplish their goals.
Experience	• A succession of events in the life of the customer dealing with an offering. This interaction will generate good or bad feelings. • Designing experiences is the new frontier in user-centered design because creating pleasant feelings is the best way to generate satisfaction, loyalty and word of mouth. • Thinking in term of customer experience forces the marketer to consider the entire sequence of events that the target customer undergoes when dealing with a company and its offerings.

Glossary

Term	Definition & Notes
Failure	• A defect, nonperformance or lack of adherence to a specific requirement. • Something that causes a product to no longer fulfill its intended function.
Failure Analysis	• A process to methodically characterize and determine the root cause of a failure. The reasoning from the symptoms of a failure to the identification of the root cause of the failure. • Used in reliability engineering, as well as FMEA (Failure Mode and Effects Analysis).
Feature Creep	• The tendency to add new features and attributes to a new product during the design process. • Feature Creep is not a desirable activity as it delays the product, adds to its complexity and may even interfere with the business objectives of the new product. It also increases the stress and aggravation of the entire development team.
Feature (a.k.a Attribute)	• A quality or capability of an offering. • When they are familiar with the product category, customers often express their needs in term of features or attributes (e.g., "I need a 4-door sedan that gets more than 25 mpg") making it more difficult to understand their primary needs.
Firefighting	• The act of addressing too many interruptions, trivial issues and other distraction at the expense of the schedule. • Taking care of surprises and other unexpected distractions. • Firefighting is generally caused by a severe lack of planning and the lack of a risk management plan.
FMEA (Failure Mode and Effects Analysis)	• A structured process to identify potential failures of new products, components or subsystems. A form of risk management. • Can also be used to analyze and determine the root cause of actual failures.
Focus Groups	• A market research tool where a qualified facilitator (moderator) leads a discussion with a select group of target customers for the purpose of understanding their needs and preferences. • Focus groups are generally eight to twelve customers in a single room; generally the room is designed for focus groups. The facilitator (moderator) leads the conversation, and the response is observed and also video taped for later analysis. • Focus groups are a qualitative form of market research.
Functional Test	• Testing a circuit board to identify functional level faults or defects.
Funnel	• See "Roadmap."

Glossary

Term	Definition & Notes
Fuzzy Front End	• The beginning of a program or project when much is not known and a high level of confusion exists. It's also the best opportunity to influence the long term success of the program or project. All programs and projects have this phase. • Initially, this term was defined as the period of time when the development of a new technology was not clear (fuzzy). The meaning has been extended to include the start of a product development project when much confusion exists — more questions than answers. • We suggest that every development project has a fuzzy-front-end; consequently, it is best to acknowledge it and set about exiting this phase by obtaining the missing information
Gantt Chart	• A horizontal bar chart used in project scheduling that shows the start and end dates of tasks within a project. • A Gantt Chart is a very common way to depict project schedules due to its simplicity and visual impact.
Gate	• The decision point, generally a meeting, at which a management decision is made to allow the product development project to proceed to the next gate. • A gate has only three outcomes: "YES", "NO" or "Delay." • Gates are generally related to the timeline of the project; consequently a gate has a chronological impact. • The metaphor of a gate is part of the Gate Phase approach to project management, which has been part of product development process for many years. This approach is now more likely replace by "Agile" and other forms of iterative product development.
Gate Phase	• A product development process or methodology based on specific stages, where at the end of one stage and prior to the start of the subsequent stage there is a review to ensure that a set of criteria has been met.
Gatekeeper	• The person who or organization that holds approval. • In product development, the person who or organization that decides if the product development team can proceed with a new stage if the criteria from the previous stage has been met.
Go/No Go Gate	• A gate in the development process that allows only two outcomes: proceed or stop. • Critical in any development process, as it determines if the development meets the criteria to proceed to the next gate in the process. • A "no go" decision does not necessarily mean end of the development project.
Goal	• A purpose toward which a project is aimed. • A general direction of achievement that is not measurable. Rather, it is statement of intent; for example, "increase sales" or "accelerate product development." • We distinguish between "goal" (immeasurable achievement) and "objective" (measurable achievement). Some choose to reverse these terms.
Good Practices	• Obtained by establishing benchmarks to identify what meets a criteria of excellence. • See also "Best Practices."

Glossary

Term	Definition & Notes
High-Performance Team	• A small number of people with complementary skills who are committed to a common purpose, performance goals and approach for which they hold themselves mutually accountable — and are deeply committed to one another's personal growth and success. — from The Wisdom of Teams, by Katzenback.
Hurdle Rate	• The minimum return on investment or internal rate of return a new product must meet or exceed as it is considered for approval and proceeds through development. • A Hurdle rate can be applied to any investment.
Ideation	• Similar to brainstorming but focused on generating specific ideas about improving the performance of a product or service. • The process of idea generation generally occurs at the early stages of development.
Inbound Marketing	• The collection of all activities that provide information about the market, including market research, customer research, feedback from other organizations like sales, training and support, industry reports, competitive analysis and technology assessments. • The function responsible for obtaining and prioritizing customer requirements, then collaborating with engineering to develop innovative solutions. • Performing both tactical competitive analysis (product-to-product) and strategic competitive analysis (company-to-company).
Infant Mortality	• Failure of a new product at the early stages of use by the customer. • It can be a metric of the rate of early failure of a new product. • Not to be confused with DOA (Dead on Arrival), which is a different metric.
Innovation	• The introduction of new capabilities that create value for customers. In the business world, innovation is so only if the customer says it is — only if the customer sees value in your offering. • The ability to deliver new value to customers across the entire set of customer experiences — from sales, to delivery, to support, operations and beyond. Every "touch point" with customers provides ample opportunity for innovation and thus value creation. • It's one thing to think up an innovation; it's another to develop it and deliver it to the customer. • Innovation is clearly not limited to technology!
Intangibles	• Attributes of an offering that can't be perceived by the senses. Goodwill, brand, image and position are examples of intangibles. • The role of intangibles in the customer decision process is often underestimated leading to products that are perceived by the customers as incomplete.
Intellectual Property (IP)	• Information, formulas, designs and other knowledge that is unique to an individual or company and can be used to grow a business or improve the competitiveness of an organization. • IP takes many forms, but the main criterion is that it can provide a competitive advantage or can be sold to someone who values it.

Glossary

Term	Definition & Notes
Issue	• A known obstacle, problem or anything that has 100% probability of occurring. And when it does occur, it will cause serious or fatal disruption to a program or project.
Kaizen	• A Japanese term that represents the philosophy of continuous improvement. • More than a set of tools or processes, Kaizen is a state or mind, a commitment to continually improve and striving for perfection.
Kano, Kano Analysis	• A model developed by Dr. Noriaki Kano to characterize quality as perceived by the customer. • Generally shown as a two-dimension diagram, the model maps the degree to which the new product performs versus the degree of satisfaction of the customer. • The model assumes that all input shown was provided by the customers, and not internal personnel. • The model can be very tedious and involved.
Lagging Metrics	• See "Outcome Metrics."
Lead Customers	• Carefully selected representatives of customer groupings (segments) that have needs and requirements typical of the rest of the segment. • In some companies, Lead Customers represent customers who are facing problems ahead of the rest of the segment, providing early visibility of future customer needs.
Leading Metrics	• See "Process Metrics."
Lessons Learned	• The benefit of performing retrospects (postmortems) and implementing the improvements as a result of the process. • Lessons learned do not benefit the organization until they are applied to improve the future performance of the organization.
Life Cycle Costs	• The total costs that the customer will incur during the useful life of the new product. These costs include purchase, maintenance, repair, installations, training operations and disposal. • Generally used during the selling cycle. Customer wants to know the total cost of a new product, not just the selling price.
Market Segment	• A group of customers who display similar characteristics relevant to a business. • Market segmentation should answer three questions: – Who are the customers? – Where are they located? – What are they trying to achieve?
Market Segmentation	• The process of grouping customers with similar characteristics. • Generally the segments represent the view of the customers; that is, how they classify themselves. • A market-driven segmentation should include your products or service. It should focus on how customers group themselves. Subsequent versions of the segmentation may include your products.

Glossary

Term	Definition & Notes
Market Sensing	• Research, observations and experience develop a deep understanding of the forces acting on a market to enable the development of a vision of the future. • Sensing the market is generally the responsibility of the senior staff and involves deep participation in the affairs of a market place. • Gathering customer needs and customer requirements is not part of Market Sensing.
Market Share	• A company's sales in a product area expressed as a percent of the total sales into that market by all competitors • Market share depends on what the company defines as its market and its competitors. • If it only considers competitors as companies that offer the same product type, it will have a higher market share than if it defines its competition as including substitution products.
Market-Driven	• A commitment to serve the needs of a market. • A chosen strategy and cultural shift that enable the achievement of a set of business objectives by focusing on serving specific market segments, gaining profound insight into specific market segments, delivering superior value and earning customer loyalty.
Marketing Requirements Document (MRD)	• The process that the marketing function uses to communicate to engineering (R&D) the problems that a new product must address. This takes the form of a document called the MRD, which articulates the prioritized customer requirements. Engineering uses the MRD to generate product concepts, features and specifications • The MRD goes by various names depending on the company or industry. There is always a document that serves to communicate the prioritized customer requirements to the R&D organization.
Mean Time Between Failures (MTBF)	• The number of failures in a given period of time divided by the total uptime in the same period. • A measure of reliability that represents the average time between failures.
Mean Time to Repair (MTTR)	• The total estimated downtime of the new product for a given period divided by the number of failures or repairs performed in the same period. • Used to plan maintenance programs, calculate Total Cost of Ownership and compare reliability.
Metrics	• A standard of measurement. For example, metrics will directly measure our progress toward achieving the organization mission and group's objectives. • There are three kinds of metrics: 1. Results metrics. 2. Process metrics. 3. Customer satisfaction metrics.

Glossary

Term	Definition & Notes
Mission	• The specific task or assignment adopted by an entire group or team. The Mission describes what it does, for whom it does it and the reason it does it. • The mission statement should be specific, achievable, easy to articulate and motivating. For example: – The mission of the police is to protect and to serve. – The mission of the product team is to get a specified product out on time and meet customer requirements.
Need	• Describes what the customer says he is lacking, what he would like to do or obtain or what he wants to get. • Customer needs, as expressed by customers, are often limited to product "features" (attributes) or problems that may or may not be related to the customer end goals or to how the decision will actually be made between competing offerings. It is the reason we introduced the term "Compelling Reason to Act."
Net Present Value	• Discounting future cash outflows and inflows using today's "cost of money." • A way to normalize several investments so you can compare apples to apples. • A way to determine the viability of an investment objectively. • Used to evaluate several alternative new products and compare their financial viability.
New Product Development (NPD)	• The processes, models and tools to develop a new product. • NPD ensures speed, repeatability and concurrency. • The processes must be documented, measured and improved in order to yield long-term benefits.
New Product Introduction (NPI)	• The process of rolling out or launching a new product to the market. • The go-to-market strategy and its corresponding tactics.
Objective	• A specific, definable and measurable end-result, which is to be pursued by a team or an individual of a company within a specific period of time. • Objective and Goal are often used interchangeably. See this definition.
Offering	• A set of products, services and intangibles offered to satisfy particular customer problems. • Express what the product is from the customer's perspective. • See also "Complete Product" and "Whole Product."
Outbound Marketing	• The function that leads all activities that take the products, programs and messages to the market. • Includes managing the distribution channels, enabling the direct sales channel in all its forms, marketing communications and customer retention programs, pricing and pricing schedules, new product roll-outs, etc.
Outcome	• The measurable outputs of an action project or program. The measurement of the actual results. • It is important not to confuse "completion" of a task with "results"; they are not necessarily the same.

Glossary

Term	Definition & Notes
Outcome Metrics	• Also called "Lagging Metrics," they track the final outcome or result of a project or process.
Pipeline	• See "Roadmap."
Plan	• A detailed formulation of a program of action; for example, a Marketing Plan. • A written set of tasks, schedules and resources with defined end dates and an assessment of any barriers which might put the schedule at risk. • There are many outlines for a plan, but generally a plan answers seven critical questions: 1. What needs to be done? 2. Who will do it? 3. By when will it be done? 4. How much will it cost? 5. What are the threats and risks? 6. Why are we doing it? 7. How will we mange change?
Policy	• A definite course or method of action to guide and determine present and future decisions. • An envelope for action with clearly defined boundaries. • The organization may choose to make policy inviolate to provide a more robust framework for action.
Portfolio Management	• A process to prioritize development projects based on a set of documented criteria. • Can be used any place where a business decision must be made. • The intent is to maximize the value of the portfolio to the corporation and to ensure that corporate strategies are in full alignment with the flow of new products. • Other names include Pipeline Management, Product Pipeline, The Product Funnels and The Product Portfolio.
Postmortem	• A process to understand the effectiveness of a project and to use the information to improve future performance. • Postmortems can be done at any time during the life of a project; they are most commonly done at the conclusion. • Postmortems are only useful if the information is used to improve the future performance of the organizations. • Postmortems are also known as Debriefs, Retrospects and Lessons Learned.
Procedure	• A written series of steps followed in a regular manner to accomplish a specific task in reproducible manner. • Similar to a process, but generally assigned to a single individual. Some compare a task with an action item.
Process	• A documented, sequential list of value-added steps that can replicate a product or service to a customer. • Unlike a project, processes are designed to be repeatable. A project is a single event with a clear start and end.

Glossary

Term	Definition & Notes
Process Metrics	• Also called "Leading Metrics," they measure activities that lead to a result. They are important to track progress towards a goal. These metrics are helpful to improve processes.
Process Out of Control	• When the number of defects exceeds boundary conditions. • Thus, an engineering organization, as a process, is out of control when there are no metrics to detect the number of defects nor a system for remedial and corrective action.
Product	• Anything that can be offered to a market that represents value; i.e., demand. • Good or service that meets the requirements of a market segment. • Colloquially, a manufactured good or the tangible part of value. But this is a limiting connotation, as customers in a business-to-business environment buy solutions to problems, and such solutions are made up of tangibles, services, experiences and intangibles. Sometimes the term "whole product" is used when referring to the entire package of tangibles, services, experiences and intangibles offered to a market.
Product Benefit	• The positive outcomes or results that an offering promises a customer. • Anticipated benefits apply before the purchase. Actual benefits apply after the purchase and manifest themselves as the customer's experience with your offering.
Product Development Team	• All the individuals whose contribution will result in the delivery of a new product to the marketplace. • In some organizations the Product Development Team is defined more narrowly; for example, only the marketing and engineering professionals involved.
Product Funnel	• See "Pipeline."
Product Life Cycle	• The four stages that all new products go through from introduction to end-of-life: introduction, growth, maturity and decline. • Duration of time from the earliest proposal to removal from the order entry system (phase out). A series of phases products follow from earliest conception to end-of-life • Product Life Cycle has many applications in product companies; such as tracking the product portfolio, ensuring profitability (gross margin time number of units), resource allocation, etc.

Glossary

Term	Definition & Notes
Product Management	• The process and strategy to mange the life cycle of a product such that it meets profitability targets. • The role of Product Management is to define, develop, deploy and maintain products and services that: – Provide more value than the competition. – Help build a sustainable competitive advantage. – Deliver financial benefit to the business. • Six of the key disciplines of product management are: 1. Market Research 2. Product Definition and Design 3. Project Management 4. Evangelize the Product 5. Product Marketing 6. Product Life Cycle Management
Product Manager	• The person assigned the responsibility and accountability for leading all the various activities that concern a particular product along its life cycle. • Sometimes "Brand Manager", particularly in consumer products. • Generally the Product Manager has P&L (Profit and Loss) responsibility for the product line. Some companies do not give P&L responsibility, but assign a "pro-forma" P&L to signify the importance of the job. • The Product Manager is responsible and accountable for achieving the business objectives of the product line.
Product Portfolio	• See "Pipeline."
Product Requirements Document (PRD)	• See "Marketing Requirements Document."
Program Evaluation Review Technique (PERT)	• A form of network analysis used to estimate the duration of a project.
Program Management	• The business function that assures the clean execution of corporate strategies, where a program is a group of related projects managed in a coordinated way to obtain the benefits and control not available from managing them individually. • The investment in program management is made to: – Reduce the variability of return from an investment. – Ensure alignment between strategy and execution in order to meet the business objectives of the enterprise. – Deal with change in a dynamic environment to ensure business results. • Program Management is not the same as "Project Management."
Program Manager	• A person with the skills, knowledge and abilities to lead a cross-functional team in the development of a product with a successful outcome such as profitable time-to-market. • This definition applies to the context of product development — your organization may use this title differently within a different context.

Glossary

Term	Definition & Notes
Programmatic Customer Visits	• A structured set of customer interviews where the participants share a common set of topics to be discussed. Further, the participants agree to a regular schedule of trips to the field to conduct the interviews. • A common way to capture, prioritize and archive the information from the customers.
Qualify Function Deployment (QFD)	• Quality Function Deployment is a structured methodology that uses a set of matrices for linking market requirements to the entire product development effort. • QFD will not define a product for you; it is simply a tool to manage information about the new product so that the integrity of the customer requirements is not jeopardized. • QFD is also called the House of Quality due to the appearance of the matrices. • QFD originated in Japan as a way to manage all the features and their interactions of a new product from start to finish.
Quality	• Perception of the degree to which a product or service meets customers' expectations. • Quality is whatever your customer says it is. • Quality is an attribute of the product that generates customer satisfaction. But it is also a perceived attribute, and therefore is somewhat subjective. • Quality Standards are approved, documented rules that define one or several dimensions of quality for a particular application or set of customers.
Rapid Prototyping	• Any of a variety of processes that avoid tooling time in producing prototypes. Generally these are nonfunctioning. • Rapid Prototyping exists to test the feasibility of a particular design with customers to reduce time-to-market.
Remedial Action	• Activities to provide an immediate resolution in order to solve a problem to the satisfaction of the customer. • Without regard to permanent solutions.
Request for Proposal (RFP)	• A process used by larger companies, government entities, military and universities to invite suppliers to bid on a specific acquisition of products, services, maintenance, etc. as a package. The submitted RFPs are then used by the buying entity to evaluate the merits and make an informed decision to purchase.
Request for Quotation (RFQ)	• See "Request for Proposal (RFP)."
Requirement (Customer)	• A brief and clear statement expressing what the customer will be able to experience as a result of owning and/or using the offering. • Product and customer requirements are two ways to express the same thing. • Requirements must be short, unambiguous, and able to be verified
Requirement (Product)	• A brief and clear statement expressing what an offering will have such that the needs of the customer will be met. • Product and customer requirements are two ways to express the same thing. • Requirements must be brief, unambiguous, and able to be verified.

Glossary

Term	Definition & Notes
Results	• The measurable achievement of an endeavor; for example, the results of a winning product is the achievement of the profitability target. • The outcome of an endeavor.
Rev (Review or Revise)	• To review or revise, used in software development to track the number of revisions that have been made to a specific application. • Revision can also be applied to documents to track version control.
Reverse Engineering	• The process of copying a product by analyzing every part of a competitor's product and then incorporating its features into one's own products. • Generally not a good thing to do.
Risk	• Something that can harm your program and should not happen. • The probability that something may go wrong with your project. • There are many types of risks that can interfere with your product development.
Risk Management	• The process to identify, prioritize, plan and manage risks in a project. • Risk Management starts at the earliest stages of a project and continues through completion.
Risk Management Plan	• The documented set of prioritized risks and corresponding mitigation plan. • The Risk Management Plan is constantly monitored.
Roadmap	• The manifestation of the product strategy for a product line or a company. • The outline of the future products versus the time it takes the organization to develop to achieve its business objectives. • Organizations may use terms like "Funnel", "Product Funnel", "Product Roadmap", "Pipeline" or "Product Portfolio." We suggest all of these terms are synonymous.
Simultaneous Engineering	• See "Concurrent Engineering"
Situation Analysis	• The current snapshot of the factors confronting a team, company or organization in a complex situation. • A short list of items, observations or data that are highly relevant to the topic at hand and are presently important to consider. • Examples include the Situation Analysis of a market and the Situation Analysis in a Product Proposal.
Sponsor, Team Sponsor. Team Mentor	• An Executive of the company who agrees to support a team in order to ensure its success. • The sponsor must have the time, willingness and skills to assist the team in a measurable way.
Strategic Alignment	• The endeavor to align all stakeholders on a single strategic direction such that their day-to-day activities fully support the endpoint. • Strategic Alignment complements (but is different from) "Strategy Formulation" or "Strategizing", which is the process of synthesizing and articulating a particular strategy.

Glossary

Term	Definition & Notes
Strategic Formulation (Strategizing)	• The act of formulating a strategy, different from the process of strategic planning. • Strategizing is the thought process or brainstorming needed to synthesize a strategy given a clear understanding of the market.
Strategic Initiative	• A program that, when completed, will deliver visible results in clear support of an overall strategy. It's defined to clearly move the organization towards the achievement of the corporate or group strategic goal. • Strategic Initiatives possess the following attributes: – Deemed critical to the success of the organization — non-negotiable and successful implementation is the only outcome. – In general, assigned to a member of the executive team as part of his MBO. – In general, implementation is carried out by one or more small teams that are given a clear charter, including deliverables, metrics, desired outcomes, time lines, roles and responsibilities. The executive is still accountable for measurable results. – Periodic reviews are included to ensure progress and accountability. – Fulfilled through programs and projects. – Span more than one fiscal year or are part of the yearly operational plan.
Strategy	• The approach that a team or company chooses to achieve a set of business objectives. • The approach guides the choices, behavior and direction of an organization.
Strategy (Corporate)	• A set of Strategic Initiatives made up of programs and projects that, when implemented, will fulfill the vision of the company. • This is the means by which you connect everyone's day-to-day activities to the vision of the company.
Strategy (Year Around Planning)	• A model of strategic planning that, unlike the yearly cycle of "business planning", assumes that strategy is a constant endeavor. • Usually characterized by a rigorous review and update of the strategy twice or three times per year.
Superior Value	• Better than your competition. Always defined by your customers, based on their comparison of you with your competitors. • Superior Value shows when a customer selects your products or solutions over that of your competition.
Switch-Over-Cost	• A term used to describe the mental process to shift one's cognitive resources from one topic to a completely different one. • Research done by Pattison, Kermit and Worker in "The Cost of Task Switching" shows that it may take up to 23 minutes to return to the same level of concentration after an interruption.
Team Charter	• See "Charter."
Team Sponsor, Team Mentor	• See "Sponsor."

Glossary

Term	Definition & Notes
Theory of Inventive Problem Solving (TRIZ)	• A Russian acronym that refers to a methodology for solving a difficult problem whereby multiple alternative solutions are defined. • The method is supposed to improve creativity by enabling the team to think of many solutions to a single problem. Generally a multi-discipline methodology to tap of the creativity of different individuals.
Timeline	• The schedule or the amount of time plus the activities needed to deliver a set of results. • A Gantt chart showing the chronological progression to the completion of a project.
Time-to-Market	• A measure of the speed by which a new product is developed. • Measured from the official start of the development until the product is ready to be purchased by a customer. • A critical measure to ensure competitiveness and success in the market place. • Various companies use different start and end points to calculate their time-to-market.
Time-To-Results (Time-To-Profit or Time-To-Cash)	• That point in the product life cycle where a target result that determines the success of the product has been met. • The success of a product may be measured by margin, revenue, market share or other strategic goals. • The time-to-results point is determined no later than the creation of the product proposal.
Total Cost of Ownership(TCO)	• The aggregate of all the costs during the life of a product, from acquisition to disposal.
Use Case	• Describes a specific activity or set of activities performed by a customer with a device or product. • Use cases are created with specific goals to understand the interaction between an operator and the equipment.
Value	• The customer's perception of what he wants to happen in a specific use and situation as the result of using a company's offering. • There is value in using the product or service — i.e., deriving a benefit as a result of use. There is also value in owning the product or service — i.e., deriving an emotional benefit just from owning the product or service.
Value Proposition	• A statement describing what resulting experience the customer will have, articulated in a way that communicates clearly and briefly the benefits, costs and competitive advantage of a particular offering. • The value proposition must be compelling and differentiated as perceived by the customer. • The resulting experience includes the outcomes or results that the customer will experience. • Whole products deliver value propositions.

Glossary

Term	Definition & Notes
Values	• A set of deeply held beliefs that enable employees to act without the need to seek management approval. • Values are never intentionally compromised. • A common set of social principles that guide the behavior and decisions of each person and the organization.
Vision	• The vivid description of the end destination and results of a group effort. • The distant lighthouse that provides direction to the completion of the team's journey. • A vision statement evokes concrete images. It is compelling and prompts forward progress. It is succinct and memorable.
Voice of the Customer (VOC)	• A process that uses structured, in-depth interviews for eliciting customer requirements. • VOC is also used as a generic way to talk about customer needs or customer requirements. • VOC by itself does not yield customer needs; other processes have to be used in conjunction to process the input from the customer, turn it into product features and eventually into a new product.
Voices	• Verbatim quotes from customers that provide insights as to their needs. • These voices are processed to extract requirements.
Whole Product	• The set of product, services, ancillary parts and intangibles needed to ensure that the customer will achieve his compelling reason to act (CRTA). • The actual product is a subset of the whole product. The whole product is what the customers buy. • Ignoring dimensions of the whole product explains why so many companies fail to penetrate new markets. • See also "Complete Product" and "Offering."

FMEA Process Map from SEMATECH, Inc.

Overview

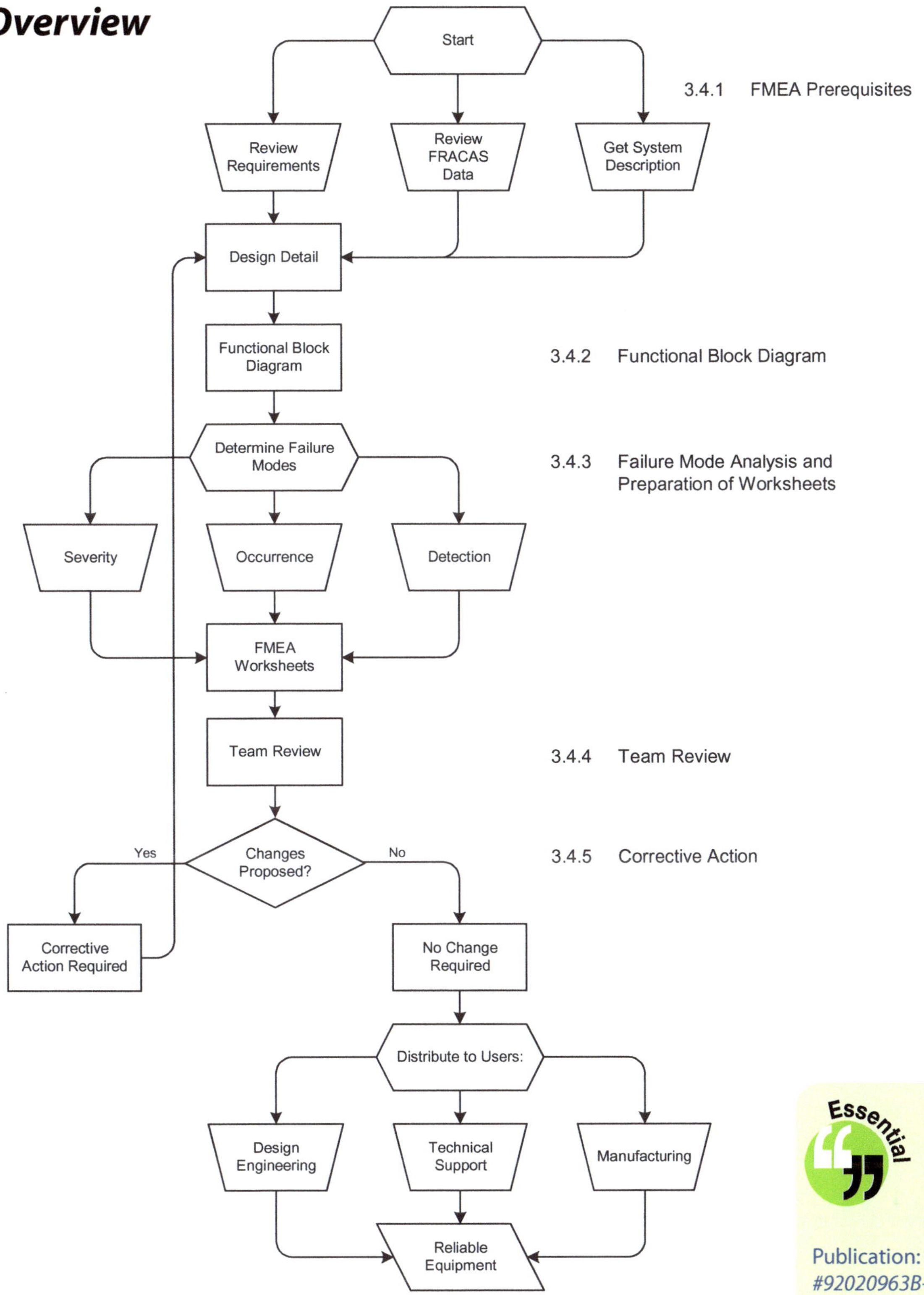

Figure 3 FMEA Process

Technology Transfer #92020963B-ENG

Essential

Reproduced from International SEMATECH, Inc., which owns all copyrights.

Publication: *Technology Transfer #92020963B-ENG*, p. 14

http://ismi.sematech.org/docubase/document/0963beng.pdf

Templates

APPENDIX D

Instructions

These templates are for you to use and share. Throughout this guidebook, you have seen real world tools and examples. Any page with a template icon indicates that there is a correlating fillable form in this section.

For more resources and future updates, visit us at: www.rapidinnovation.com.

Templates with Examples

Example: Types of Risk Management 39

Charter for a Risk Management Team 47

Documenting Your Risk Management Plan. 71

- Approach 1: Traditional Table 71
- Approach 2: Electronic Spreadsheet 71

Monitor Your Risk Plan Using the Critical Questions of Proactivity 74

Risk Management Strategy Matrix for a Development Program 82

- Example 1: Types of FMEA (Y) Versus Program and Project Levels (X) 82
- Example 2: Types of FMEA (Y) Versus the Design Phase (X) 83

The FMEA Management Plan 94

- Example 1: Basic FMEA Management Plan 94
- Example 2: Expanded FMEA Management Plan 96

Planning Checklist 111

Planning Document 112

More Tips for Users of the Interactive PDF Version

You can view the complete list of templates with the "Show/Hide Templates" bookmark on the left Navigation Panel. This reveals the Attachments Tab at the bottom of the screen. Double click any template you want to work on. You can also open a template by double clicking the white pushpin icon on the specially marked example pages.

Once opened, these templates are enabled for use in Adobe Reader or Adobe Acrobat — templates that you can fill with your own information and share with your peers and colleagues.

When you open original templates from your guidebook, they will be fresh. But you can save versions for future access — whether you have completed the template, partially filled in your information or left in blank. Saved versions remain enabled for revisions and collaboration.

When you close the template, you will return to this guidebook.

These enabled templates also provide you with Comment & Mark-up Tools. Just as you might with a hard copy, you and your team can highlight text, add notes, and make custom annotations.

Types of Risk Management to be Performed

	Level of the Risk Management Plan			
Types	**Program Level**	**System Level**	**Subsystem Level (Module)**	**Component Level**
Financial (Business case)				
Marketing (Market)				
Design (Software, hardware)				
Manufacturing (Process)				
Supply Chain (Suppliers)				
Service (Service and repairs)				

Charter for a Risk Management Team

CHARTER FOR: ______________________________

THE BUSINESS CASE

____

____

____

____

____

____

THE CASE FOR COMPLEXITY

____

____

____

____

____

____

____

____

THE MAJOR CONTRACTORS

____

____

____

____

____

____

OBJECTIVE FOR RISK MANAGEMENT

____

____

____

____

____

____

OVERALL RISK MANAGEMENT STRATEGY

____

____

____

____

____

____

____

____

____

____

____

BUDGET

____

____

____

Charter for a Risk Management Team (Cont'd)

Leadership

Management Sponsor ______________________

____

Overall Program Manager ______________________

____

Product Quality Assurance ______________________

____

Core Team Leader ______________________

____

Core Team ______________________

____

____

____

____

Overall Facilitator

____

____

Critical Success Factors

____

____

____

____

____

____

Decision-Making

____

____

____

____

Key Assumptions

____

____

____

____

____

____

____

____

Key Constraints

____

____

____

____

____

____

____

____

____

Contact the author: risk@rapidinnovation.com.

Documenting Your Risk Management Plan

Template 1

Risk Management Plan

Work Package or Task	Risk Event No. and Description	Probability	Impact	Plan			Actual			Owner
				Risk Event Status (P x I)	Criticality	Impact on Other Risks	Preventive	Trigger Point	Contingent	

Documenting Your Risk Management Plan

Template 2

Risk Management Plan

Work Package or Task	Risk Event No. and Description	Owner	Criticality	Plan			Actual			
				Preventive	Trigger Point	Contingent	Status	Date of Impact	Risk Impact	Recovery Action

Contact the author: risk@rapidinnovation.com.

Documenting Your Risk Management Plan

Template 3

Risk Management Plan

				Plan			Actual			
Cause of Risk	Description of Risk	Description of Impact	Probability that Risk Will Happen	Impact of Risk if it Does Happen	Ranking Based on P and I	Trigger Point	Contingency Plan A	Contingency Plan B	Risk Owner	Comments

Examples: Documenting Your Risk Management Plan

Example for Template 1

Work Package or Task	Risk Event No. and Description	Probability	Impact	Plan			Actual			Owner
				Risk Event Status (P x I)	Criticality	Impact on Other Risks	Preventive	Trigger Point	Contingent	
Develop blueprints	Draftperson not available due to illness	2	7	14	Critical	Yes Engineering	Hire another draftperson	Draftperson calls in sick	Borrow draftperson from another department	Fred Smith
Stone work completed	Stone mason from Europe unable to arrive on time	5	6	30	Critical	No	Coordinate travel arrangements well in advance	The stone mason is not on site 6 days prior to start	Hire local stone mason to start positioning the material	May Jones

Example for Template 2

Work Package or Task	Risk Event No. and Description	Owner	Criticality	Plan			Actual			
				Preventive	Trigger Point	Contingent	Status	Date of Impact	Risk Impact	Recovery Action
Develop Blueprints	Draftperson not available due to illness	Marsha L.	C (Critical)	Hire another draftperson	Draftperson calls in sick	Borrow draftperson from another department	Draftperson has announced her resignation	September 2, 2005	Two-week delay	Activate preventive (hire)/Stage contingent (borrow)

Example for Template 3

Cause of Risk	Description of Risk	Description of Impact	Probability the Risk Will Happen	Plan			Actual			
				Impact of Risk if it Does Happen	Ranking Based on P and I	Trigger Point	Contingency Plan A	Contingency Plan B	Risk Owner	Comments
Draft-person is not available due to illness	Causes a delay in the creation of the schematics	Delays entire project by the same number of days	60%	$50,000 loss	#5	Draft-person calls in sick	Borrow a draft-person from another division	Hire a temporary draft-person	Fred A.	Will need to train either replacement

Monitor Your Risk Plan Using the Critical Questions of Proactivity

This list can also be used as the agenda for your Risk Management Review meeting.

What has changed since we last met?

- ❑ What important changes have taken place that impact our project?
- ❑ Are there any new risks that must be added to the Risk Management Plan?
- ❑ Who are the risk owners for the newly identified risks?
- ❑ Are there any old risks that are obsolete and must be deleted from the Risk Management Plan
- ❑ What changes have happened to the current list of risks?
- ❑ What other changes have take place?

What are the reasons for the change or variations?

- ❑ What has caused the new risks?
- ❑ What caused the changes to existing risks?
- ❑ What cause the deletion of risks?
- ❑ What are the implications of the changes?
- ❑ Are there any trends that will impact our Risk Management Plan?
- ❑ What will happen if the new trends continue?
- ❑ What should we do right away to mitigate the trends?

What are the specific changes to our current Risk Management Plan?

- ❑ Add new risks?
- ❑ Delete risks?
- ❑ Changes to current risks?
- ❑ Changes in responsibilities?

Risk Management Strategy Matrix for a Development Program

Template 1: Types of FMEA (Y) Versus Program and Project Levels (X)

Types of FMEA	Program Level	Project Level
Financial (Business Case)		
Marketing (The Market Place)		
Engineering Design (Software and Hardware)		
Manufacturing (Process)		
Supply Chain (Suppliers)		
Service and Repair (Field service, repair and maintenance)		

Risk Management Strategy Matrix for a Development Program

Template 2: Types of FMEA (Y) Versus the Design Phase (X)

Each of the cells in the matrix is an opportunity for FMEA — use your judgment, guided by the business objectives of your project, to determine the number, type and timing of your FMEA.

	Level of the Risk Management Plan During the Product Life Cycle				
Types of FMEA	**Product Definition**	**Product Design**	**Product Rollout**	**Sustaining Phase**	**Pre-Phaseout**
System level					
Subsystem level					
Component level					
Nano level					
Process FMEA					
Software level					
Service level					

FMEA Management Plan

Template 1: Basic

FMEA Management Plan										
Tracking Number	Statement of Failure (Risk)	Possible Root Cause	Statement of Impact	Occurrence (Probability)	Severity (Impact)	Detectability	RPN	Preventive Action	Mitigation	Owner

FMEA Management Plan

Template 2: Expanded

FMEA Management Plan											
Tracking Number	Component Description	Failure Mode (Risk Description)	Occurrence (Probability)	Severity (Impact)	Detectability	RPN	Mitigation	Trigger	Risk Owner	Date of Last Review	Status at Last Review

Planning Checklist for Deploying a New Tool

Deploying a tool is indeed a project. Consequently, ensure that you have a deployment plan, a schedule, budget, etc. Working with your Core Team, or with the peers who have agreed to help, you can develop a tactical plan that will ensure a successful deployment.

The list below is intended to help you formulate your own plan. You may need to add your own items to the list. Also note that the list below is in no particular sequence. The sequence will be determined by your specific needs and schedule. Finally, the list is not rigid. Use you judgment and do what makes sense; you know your organization better than most.

Planning Checklist

- ❑ Schedule for the deployment defined, documented and approved.
- ❑ Training materials for employees developed and documented.
- ❑ Training materials produced (hard & soft copies), available for use.
- ❑ Training materials placed in the intranet or Web page.
- ❑ Trainers on the new tool or method identified and informed of their duties.
- ❑ Training of employees scheduled and employees notified.
- ❑ Training of managers and supervisors scheduled.
- ❑ Feedback mechanism in place.
- ❑ Metrics to measure the progress and eventual success of the deployment identified and ready to use.
- ❑ Coaching and hand-holding for managers, employees and stakeholders defined and ready — including those who will do the coaching.
- ❑ Support (assistance) available; for example, facilitators, coaches, tech support, etc.
- ❑ Obstacle removal in place.
- ❑ Templates, checklists and other tools ready and available.
- ❑ Information on our Web page done.
- ❑ Processes that will be impacted by the deployment identified and corrected.
- ❑ Changes to our quality manual completed.
- ❑ Manuals, guides and other reference materials that will be needed by the users ready and available.
- ❑ Executives briefed and ready to fully support the deployment.
- ❑ Human resources (HR) informed.
- ❑ Training department informed.

Planning Document for Deploying a New Tool

This tool may help you secure assistance from your executives. It is also a good point to start the process of deployment. It may even become a standard template for deployment of future tools and templates. You will want to modify it to meet your specific objectives and adapt to the culture of your organization.

1. Tool or methodology to be deployed

2. Objective of the tool deployment

3. Who is the "expert" on this tool?

4. What documentation is available?

5. Are there any external resources available?

6. What are the risks involved in the deployment?

7. How will we "pilot" the tool?

8. What will be our communication plan?

9. What will we measure to assure us that the deployment has been successful?

FREE BONUS

How to Get Your Interactive Version of this Guidebook

Register Your Book

Thank you for purchasing *Risk Management and FMEA for Product Development*. Many find the interactive PDF version of this guidebook to be a valuable addition because the cross-reference links are live and all the templates can be filled out electronically and used to collaborate with your team members around the world. Once you enter the information, every template becomes a separate PDF file for sharing and collaboration. Furthermore, the electronic version of this guidebook can also be stored in any portable device that accepts PDF files such as tablets, laptops and even smartphones.

To obtain your free interactive PDF copy of *Risk Management and FMEA for Product Development*, follow these simple steps:

1. Send an email to risk@rapidinnovation.com with "**Registering RISK book**" in the subject line. In the body of your email, include your name plus the following registration code: **RSK4712E**.
2. In a few days, you will receive an email notice that your book has been registered with either an attached copy of your free interactive PDF or a link to download your copy.

If you have any questions during registration, please email the publisher at: risk@rapidinnovation.com.

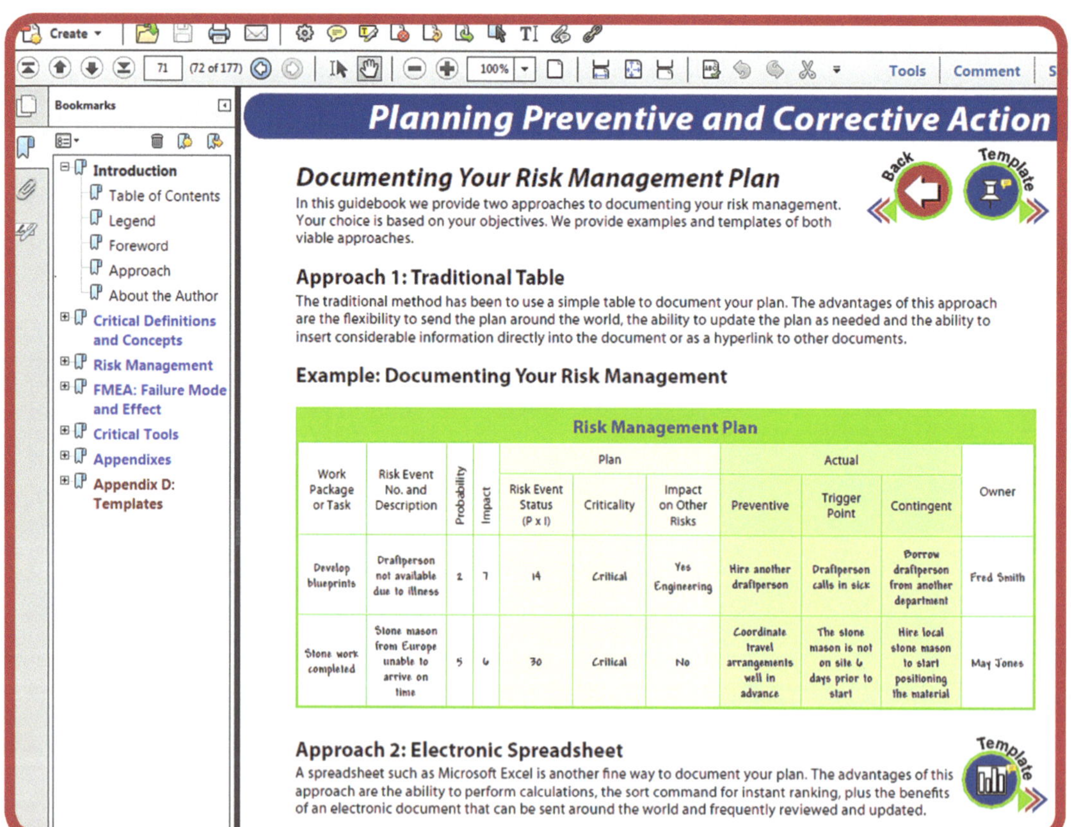

Other Titles You May Be Interested in Owning

THE SERIES

Flexible Project Management for Product Development

Traditional project management approaches are not sufficient for product development due to the complexity of getting a new product to market. Technology changes, personnel turnover, changing customer requirements, multi-site development, and unclear schedules are but a few of the variables that render conventional project management deficient.

Flexible Project Management for Product Development allows you and your team to manage your development project in the middle of the turmoil created by global competition. Beyond just reading very useful material, you will be able to use the templates and tools and also be able to share the information with your peers around the world. This is a step-by-step how-to guide. The authors realize that in the real world, you need more guidance and enough detail to enable you to apply this material to your specific development project. It starts by helping you to organize your development project to ensure that you build a robust environment to deal with chaos.

The Voice of the Customer for Product Development

Your task is to create differentiated value for your customer in order to achieve the right gross margin at minimum COGS. This guidebook is based on experiences with many technology development teams around the world — all of them focused on achieving the right volume, margin and profitability. You will be introduced to a customer-driven methodology that has been perfected over the last 30 years.

The mission of this guidebook is to create awareness of the framework and motivation needed to successfully understand customer needs, process the information and put it to use to develop differentiated products that achieve the right level of profit. This guidebook methodically walks you through the tools, templates and methods for obtaining, processing and using customer input — like no other of its kind, it is based on real projects and it is documented to provide maximum guidance to enable application.

You will be able to increase the value of your offering offerings and create defensible market segments. You will learn how to identify opportunities by interviewing the right customers. You will also learn how to transform these opportunities into requirements and then into product concepts that cast an optimum balance between usefulness, usability and desirability. The overall objective is to create the capacity for improved productivity!

Debriefs and Postmortems for Product Development

Studies show that debriefs, or postmortems, are the best way to improve the productivity of teams — better than training, books, seminars and other tools. Your ability to methodically identify areas of improvement from development programs and immediately apply them to upcoming efforts will result in improved productivity, more enthusiastic teams and improved morale in your development organization.

This step-by-step guidebook will help you avoid the common mistakes many organizations make that block them from obtaining the full benefits from debriefs. You will learn how to prioritize improvement changes and how to deploy them through the organization in a way that it is embraced by all.

www.ingramcontent.com/pod-product-compliance
Lightning Source LLC
LaVergne TN
LVHW070951210926
848482LV00002B/182